50478397 8

AF743534

[QH 455 JOH]

...ARY SCIENCE LIBRARY

UNIVERSITY OF BRISTOL
LIBRARY

22 JUL 2024

WITHDRAWN

The Institute of Biology's
Studies in Biology no. 70

Population Cytogenetics

Bernard John
D.Sc.
Professor of Population Biology,
Research School of Biological Sciences,
Australian National University, Canberra

Edward Arnold

© Bernard John 1976

First published 1976
by Edward Arnold (Publishers) Limited
25 Hill Street, London WIX 8LL

Board edition: ISBN: 0 7131 25969
Paper edition: ISBN: 0 7131 25977

All Rights Reserved. No part of this publication may be reproduced, stored in a retrieval system, or transmitted in any form or by any means, electronic, mechanical, photocopying, recording or otherwise, without the prior permission of Edward Arnold (Publishers) Ltd.

UNIVERSITY OF BRISTOL
LIBRARY

22 JUL 2024

WITHDRAWN

Printed in Great Britain by
The Camelot Press Ltd, Southampton

UNIVERSITY
OF BRISTOL
VET. SCIENCE

General Preface to the Series

It is no longer possible for one textbook to cover the whole field of Biology and to remain sufficiently up to date. At the same time teachers and students at school, college or university need to keep abreast of recent trends and know where the most significant developments are taking place.

To meet the need for this progressive approach the Institute of Biology has for some years sponsored this series of booklets dealing with subjects specially selected by a panel of editors. The enthusiastic acceptance of the series by teachers and students at school, college and university shows the usefulness of the books in providing a clear and up-to-date coverage of topics, particularly in areas of research and changing views.

Among features of the series are the attention given to methods, the inclusion of a selected list of books for further reading and, wherever possible, suggestions for practical work.

Readers' comments will be welcomed by the author or the Education Officer of the Institute.

1976

The Institute of Biology
41 Queens Gate,
London, SW7 5HU

Preface

Most species, and especially those of wide distribution, are not uniformly distributed over the territory they occupy. Rather they are subdivided into a series of populations within which individuals associate for reproductive as well as ecological reasons. The distinctiveness of any one of these populations and the amount of variability it contains depends on the number and type of individuals from which it arose, the kinds of selection that have been operative in it since its inception, the breeding structure of its members and its fluctuations in size. In this book we shall be concerned with one component of the genetic variation present in a population, namely that relating to its cytogenetic structure.

A spectacular range of organisms has now been shown to exhibit chromosome variation both within and between populations, and what was formerly regarded as a somewhat infrequent and unusual occurrence can now be seen to be widespread and common. It is clear therefore that in population cytogenetics we are dealing with a phenomenon of considerable biological importance, and one, moreover, which has in the past been too often ignored by students of evolution.

Canberra,
1975

B. J.

Contents

1 Some Basic Genetic Principles

Ignorance of the law excuses no man: not that all men know the law but because 'tis an excuse every man will plead, and no man can tell how to refute him.

John Selden

PRINCIPLE NO. 1 The major features of heredity in all biological systems are controlled by structural elements termed chromosomes and in diploid organisms or diploid phases of organisms there are two homologous sets of chromosomes. In addition to their visible microscopic structure each chromosome possesses a genetic structure and homologous chromosomes share an equivalent genetic organization. This consists, in part, of a specific linear arrangement of genetic loci. Chromosomes also carry a series of repeated nucleotide sequences (p. 7) which do not function as conventional genetic loci.

PRINCIPLE NO. 2 Each gene locus may exist in two or more forms termed alleles and the total allelic content of an individual constitutes its genotype. The component genes within a genotype interact with one another and with the environment. The product of these interactions is termed the phenotype and Waddington has introduced the term epigenesis to describe the interactions by which the phenotype arises during development. Heredity, on the other hand, is the process whereby the epigenetic information contained within the genotype is transferred from one generation to the next.

PRINCIPLE NO. 3 Differentiation within an individual (epigenesis) and hereditary differences between individuals are both determined by the genotype. In epigenesis the expression of the genotype changes within the individual either as a result of simple switch mechanisms which regulate the differential expression of particular gene loci at particular times in particular tissues or else as a consequence of the gain or loss of whole chromosomes or parts of chromosomes by differential replication or elimination. Such epigenetic changes are not transmitted between generations but the genetic basis responsible for the control of these changes is.

PRINCIPLE NO. 4 The individual units of the genotype are also subject to permanent change as a consequence of allelic mutations which alter the phenotype in a heritable manner. Most of this change is of little value in a stable environment since mutations occur at random relative to their prospective utility. That is mutations appear regardless of whether they are, or can ever be, useful. In a changing environment, however, a mutation may acquire use. Under such circumstances the environment acts as a selecting agency which determines that the mutant leaves most progeny.

PRINCIPLE NO. 5 Selection is merely a convenient name for the interactions that go on between an organism and its environment. Some selective agents, like temperature or water availability, are features of the physical environment. Others, like crowding or predation, are effects of one organism on another and are aspects of the biotic environment. Different environments select different phenotypes and to the extent that their differences are genetically determined this means different genotypes. The diversity found in living organisms thus reflects both their adaptation to different habitats and the differing ways in which they have become adapted to the same habitat.

PRINCIPLE NO. 6 Any given generation is descended from only a small fraction of the previous generation and this depends on the unequal capacities which individuals of differing phenotype show for survival or reproduction in a particular environment. The fitness of an organism thus has two essential components:

(a) viability, which is a measure of the probability of a genotype surviving to reproductive age;

(b) fertility, which is a measure of the number of functional gametes produced by a genotype.

PRINCIPLE NO. 7 The distribution of genetic material between generations depends on the mode of distribution of chromosomes or their parts. In asexual systems the chromosome complement is perpetuated by mitosis whereas in sexual systems its continuity is determined by a combination of meiosis and fertilization. Thus whereas in asexual reproduction mitosis governs heredity, in sexual reproduction meiosis and fertilization determine both heredity and variation.

PRINCIPLE NO. 8 When particular alleles of two distinct genes are situated on the same chromosome they are said to be linked and when two such linked alleles are located close to one another they tend to appear together in the gametes more often than would be expected on a random basis. Linked alleles which are widely separated on a chromosome, however, are subject to recombination as a consequence of an exchange of genetic material between homologues at meiosis. This event is termed crossing-over and is expressed at the cytological level by the formation of chiasmata between paired homologues (see *Studies in Biology* no. 21).

PRINCIPLE NO. 9 Alleles of unlinked loci may also be subject to recombination as a consequence of the common tendency of non-homologous chromosomes to orientate, and hence assort, at random during meiosis with respect to one another. Of course neither crossing-over nor random assortment in themselves guarantee new genetic combinations. This depends on the existence of prior genetic differences between maternal and paternal genotypes in respect of their allelic content. When allelic heterozygosity is high most recombination events will produce new genetic combinations. When little allelic heterozygosity exists only a fraction of the recombination events will be genetically effective. The higher the level of allelic heterozygosity and the greater the amount of recombination the larger will the number of genetically different gametes produced in a population be. Recombination allows for mutations of different origin to be brought into combination in one zygote or even one chromosome and is the principal immediate source of genetic variability within populations.

PRINCIPLE NO. 10 The extent of recombination is influenced by factors of two kinds.

(a) External factors, including the size and structure of the population. These influence the degree of inbreeding. Thus the gametes produced by a small population are more likely to be related than those produced in a large population.

(b) Internal factors, which include two components which collectively form the genetic system:

(i) the breeding system, which determines the nature of the zygotes produced from the gametes;

(ii) the meiotic system, which determines the nature of the gametes produced from the zygotes.

While asexual reproduction may occur in a variety of ways its consequences are always the same—it gives rise to offspring which, excluding mutation, are genetically identical both with each other and with the single parent from which they arise. In sexual reproduction the relationship of the offspring to the two parents involved is a function of the genetic relationship of those parents and hence the gametes they produce. Thus, although sexual processes are only incidental to reproductive activities they are vital in controlling the variation found in living systems.

PRINCIPLE NO. 11 In animals the meiotic machinery is much more varied than in plants. But in plants the diversity of reproductive mechanisms is far greater than in animals including as it does systems of self-fertilization, which are rare in animals, and hermaphroditism on a much more extensive scale than is found among animals. In all but the simplest organisms sexual reproduction is accompanied by gametic differentiation. In addition in most animals and some plants there is also a separation of the sexes, the two kinds of gametes being produced by different unisexual zygotes. This separation of the sexes is a device for the promotion of outbreeding. There are, of course, alternative mechanisms which achieve the same purpose in plants. Thus incompatability systems (see *Studies in Biology* no. 12), protandry, protogyny and special floral structure may serve as alternatives to unisexuality.

PRINCIPLE NO. 12 The release of variation within a population may be regulated in different ways in different species because the components of the genetic system may act in either a synergistic or an antagonistic manner. In plants, for example, it is not uncommon to find higher chiasma frequencies in inbreeders as compared with related outbreeders. Here one component, the higher chiasma frequency (meiotic system), serves to increase variability whereas the other, inbreeding (breeding system), has the opposite effect. Again in a number of related plant species chiasma frequency differences (meiotic system) can be correlated with longevity of the individual (breeding system); those of annual habit have a higher chiasma frequency than those with a more perennial habit. On the other hand annuals are more often inbreeding and have fewer chromosomes than related perennials, features which may both serve to restrict the variability of populations.

2 Stability and Change in the Chromosome System

Change is not made without inconvenience, even from worse to better.

Richard Hooker

2.1 Chromosome constants

Chromosomes may be characterized by properties of two kinds.

2.1.1 Chromosome morphology

The length of a chromosome is a constant and characteristic property and chromosomes may be arbitrarily classified as long (> 10 μm), medium (4–8 μm) or short (> 2 μm). In addition, while chromosomes are known where the property of attachment to the division spindle is diffused along the length of a chromosome (p. 32), a majority of organisms possess a single site of attachment whose location is again a heritable property of that chromosome. This site is referred to as the centromere or kinetochore and it is customary to recognize three major classes of chromosomes (Fig. 2.1).

(a) Metacentric: where the centromere is median or near median so that the chromosome has two well defined arms with a length ratio varying from 1:1 to 2.5:1.

(b) Acrocentric: where the centromere is close to one end of the chromosome so that one arm is substantially smaller than the other and the arm ratio ranges from 3:1 to 10:1.

(c) Telocentric: where the centromere is a strictly terminal entity and the chromosome is one-armed.

In meta- and acro-centrics the site of location of the centromere is defined by the presence of a pronounced constriction along the length of the chromosome. Other constrictions of constant location may also be present in particular chromosomes and these secondary constrictions also provide morphological markers which serve to give particular chromosomes a distinctive morphology.

Conventionally, homologous chromosomes agree in morphology. In many animals and a few plants, however, sexuality is associated with a system of balanced chromosome polymorphism in which the two sexes are differentiated in respect of their chromosome constitution (Fig. 2–1). The simplest of these systems is that in which one sex is homozygous and the other heterozygous for a single pair of structurally differentiated chromosomes which often, but not always, retain at least partial homology as in cases of XX XY♂ (male heterogamety) or ZX♀ ZZ♂ (female

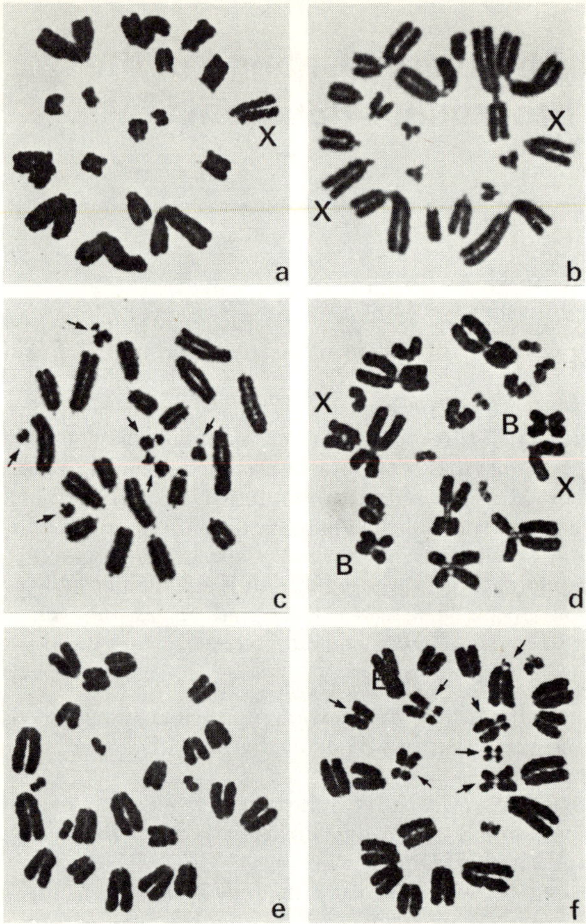

Fig. 2–1 Variation of mitotic chromosome morphology in grasshoppers. (**a**) *Chorthippus parallelus* male ($2n=17$, XO). There are six long metacentrics and eleven telocentrics ranging in size from medium to small. (**b**) *Myrmeleotettix maculatus* female ($2n=18$, XX). The complement is similar to that of *Chorthippus*. (**c**) *Chortoicetes terminifera* female ($2n=24$, XX). The large and medium chromosomes are all telocentric. Of the six small chromosomes two are telocentric, two are acrocentric and two are metacentric. (**d**) *Myrmeleotettix maculatus* female with two metacentric supernumerary chromosomes ($2n=18+2B^{m}$; compare with Fig. 2–1b). (**e**) *Cryptobothrus chrysophorus* female ($2n=24$. XX). In the standard complement all the chromosomes are telocentric. Note the presence of four small elements. (**f**) *Cryptobothrus chrysophorus* female. This individual is taken from the same population as (e) but, unlike it, is homozygous for two centric transpositions and heterozygous for a further two such transpositions involving one small and one large chromosome (arrows). In addition it is heterozygous for a supernumerary segment in the smallest autosome (arrow) so that only three small elements are present (compare with Fig. 3–2c).

heterogamety). Multiple sex chromosome systems are also known and one example of this type with X_nX_n♀ X_nY♂ mechanism will be referred to later (p. 48 and p. 59).

2.1.2 *Heterochromatic differentiation and chromosome banding*

Particular regions of chromosomes or in some cases entire chromosomes may show differential behaviour at interphase producing a visible heteropycnosis of the chromosome or chromosomes concerned. Because they are more condensed such heteropycnotic bodies also stain more deeply and so merit the term heterochromatic. Where both homologues show an equivalent heteropycnosis the heterochromatin is described as constitutive in character. By contrast in facultative heterochromatic states the two homologues differ in behaviour, only one exhibiting heteropycnosis. There is now compelling evidence that heterochromatic regions, whether constitutive or facultative in character, are genetically inactive at least in respect of any major gene loci present within the heterochromatin.

Chromosomes which have been subjected to denaturation-renaturation procedures after fixation exhibit intense Giemsa staining in regions which include constitutive heterochromatin giving rise to so-called C-bands. Such segments commonly occur next to the centromere (Fig. 2–2a). C-banding is also found at secondary constrictions and in satellite arms, i.e. short segments which extend beyond secondary constrictions.

In mammals treatment of chromosomes with quinacrine (the antimalarial drug Atebrin) produces an extensive linear differentiation of the chromosomes as a result of the presence of transverse fluorescing bands (so-called Q-bands) along the length of each chromosome. The number, size, intensity and distribution of these Q-bands is specific for each pair of homologues. Banding patterns similar to those observed with Q-fluorescence may also be obtained in chromosomes which have been exposed to a warm saline composed of a mixture of sodium chloride and trisodium citrate (SSC) for 1 hr at 60°C before staining with Giemsa. These bands are referred to as G-bands (Fig. 2–3) and similar bands can be obtained following pre-treatment with trypsin (Fig. 2–2b). While regions that fluoresce with quinacrine generally also stain with Giemsa, certain regions which show little fluorescence nevertheless stain with G-band techniques.

Three types of organization can, therefore, be defined within the mammalian chromosome with the use of banding techniques. Procentric regions characteristically show intense C-banding and some, at least, of these regions have been shown to contain sequences of highly repetitive DNA which appear to be firmly bound to acidic proteins. The remainder of the chromosome arm can be divided into band regions, which react directly with quinacrine and indirectly with Giemsa, and non-band

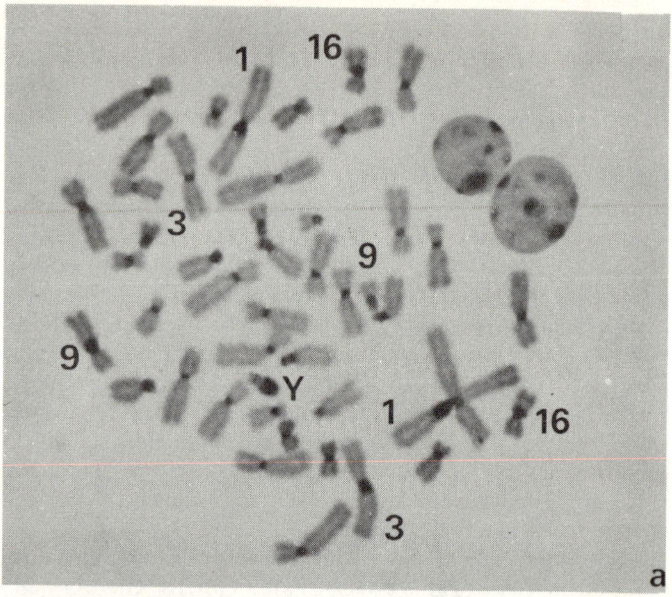

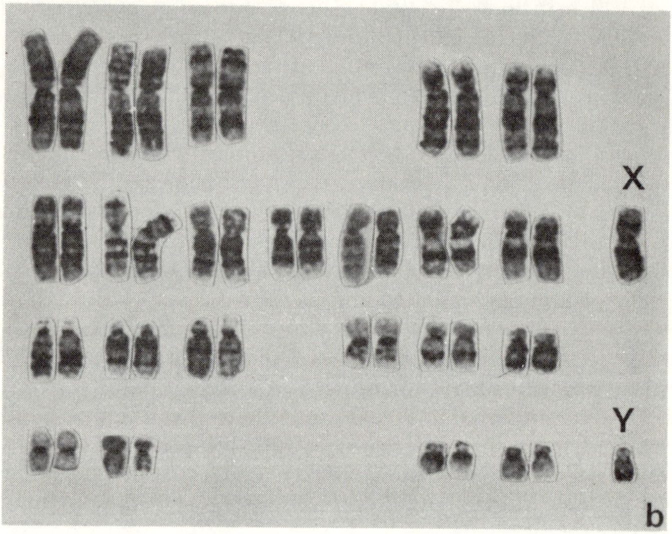

Fig. 2–2 The chromosomes of man ($2n=46$). (**a**) C-banded metaphase of a male mitotic complement with heteromorphism for the C-band expression in autosome pairs 1, 3 and 9. Note also the well differentiated Y-chromosome and the large C blocks in chromosome 16 (photograph kindly provided by Professor John Evans). (**b**) Banding of a male complement produced with trypsin treatment prior to Giemsa staining. The chromosomes have been arranged in the convention shown in Fig. 2–3 (photograph kindly provided by Dr. Marina Seabright).

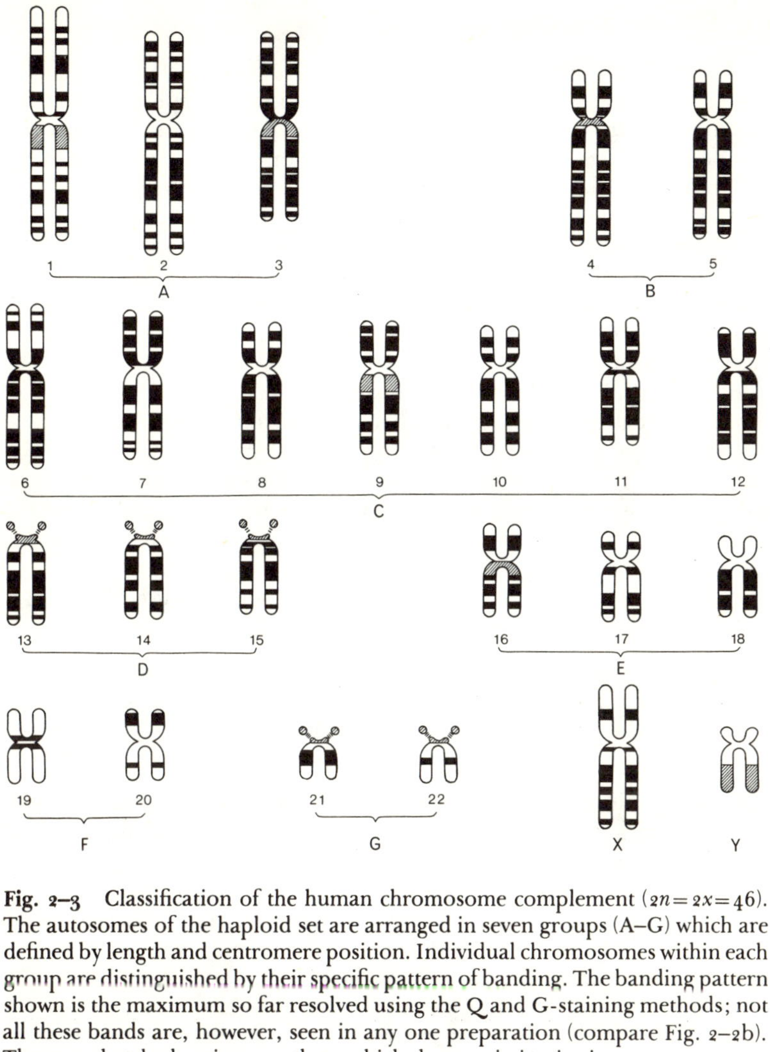

Fig. 2–3 Classification of the human chromosome complement ($2n=2x=46$). The autosomes of the haploid set are arranged in seven groups (A–G) which are defined by length and centromere position. Individual chromosomes within each group are distinguished by their specific pattern of banding. The banding pattern shown is the maximum so far resolved using the Q and G-staining methods; not all these bands are, however, seen in any one preparation (compare Fig. 2–2b). The cross-hatched regions are those which show variation in size.

regions which fail to react with either of these dyes. Finally C-band regions and Q- and G-staining regions characteristically undergo late replication in the synthetic (S) phase of the cell cycle.

In animals other than mammals, as well as in plants, Q- and G-bands are far more restricted in distribution or else are completely absent. Why this should be so is not known. It may reflect a distinctive organization of repeated nucleotide sequences within mammalian chromosomes.

2.2 Chromosome mutation

2.2.1 *Categories of change*

The karyotype may be defined as the phenotypic appearance of the chromosome complement and this, as we have seen, includes both the number and the morphology of the chromosomes. Chromosome mutations, which consist of alterations in the structure and/or the number of chromosomes, spring from causes of three kinds.

(a) Alterations, most commonly additions, in the amount of constitutive heterochromatin within a given chromosome. For example, in man variable C-band heterochromatin has been reported in the Y and in autosomes 1, 3, 4, 9, 16, 21 and 22 (Fig. 2–2a) and at least some of these variants have been shown to be heritable.

(b) Errors in chromosome movement which lead to changes in chromosome number involving either single chromosomes (aneuploidy) or else entire sets of homologues (polyploidy). Aneuploidy may involve gain (polysomy) or loss (monosomy) of chromosomes and where polyploidy is involved it may occur within a single species (autopolyploidy) or else following hybridization between two species (allopolyploidy).

It is worth noting at this point that the term haploid is used in biology in two senses:

(i) as the chromosome number of the haplophase (symbol n). Here it refers to the gametic number of chromosomes, and (ii) as the basic number of a polyploid series (symbol x). Used in this sense it means monoploid.

Thus aneuploids may be monosomic ($2n=2x-1$), trisomic ($2n=2x+1$) or tetrasomic ($2n=2x+2$) and so on. Polyploids may be triploid ($2n=3x$), tetraploid ($2n=4x$) or pentaploid ($2n=5x$) and so on.

(c) Exchanges within or between chromosomes. The mechanism of production of such exchanges is not known with certainty. It is usual to assume that it depends on breakage followed by the reunion of broken surfaces. Such exchanges can be classified in a variety of ways. We will adopt the following classification (Fig. 2–4).

(A) Those which do not lead to any alteration in either the structure or the position of the centromere

(1) Paracentric inversions

(2) Equal reciprocal translocations or interchanges

Rearrangements of this type do not alter chromosome morphology at all.

(B) Those that change the position of the centromere

(1) Unequal reciprocal translocations or whole arm exchanges.

(2) Pericentric inversions

(3) Centric shifts

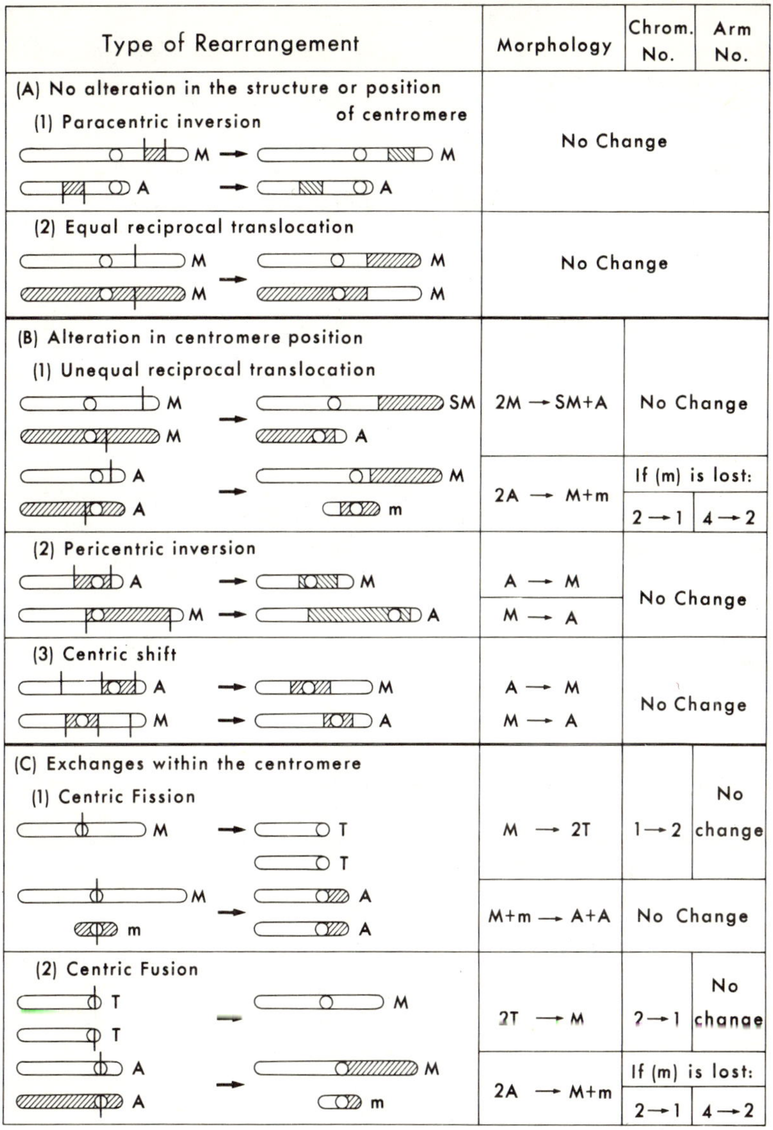

Fig. 2–4 A classification of the most common exchange types leading to structural rearrangement of the karyotype. This diagram is meant to be representative rather than comprehensive and covers all the rearrangements subsequently referred to in the text. Note that centric transpositions are usually attributed to pericentric inversion when detected through karyotype analysis (for example p. 39). The distinction between a centric shift and a pericentric inversion can, however, usually be made with confidence only after meiotic analysis.

Pericentric inversions and centric shifts alter chromosome morphology but do not alter either the number of chromosomes or the number of chromosome arms. Only whole arm transpositions involve an alteration in chromosome number and only then in cases where the small centric element is lost. Where such a loss does occur there is a simultaneous reduction in the number of chromosome arms.

(C) Those which involve exchange within the centromere

(1) Centric fission

(2) Centric fusion

Both categories of centric exchange alter chromosome morphology and chromosome number but do not change the number of chromosome arms except in the case of loss of the small centric element following strict centric fusion between acrocentric elements.

2.2.2 *Limiting factors*

(a) THE PRINCIPLE OF BALANCE The haploid complement of chromosomes contains one complete set of genes which functions as an integrated epigenetic system. In both plants and animals the capacity of a zygote to develop is seriously impaired by departures from a balanced genome containing two or more entire sets of genes. Loss is more serious than gain though both alter the dosage of genes and lead to some disturbance in the pattern of development. Loss from diploids is more serious than loss from polyploids which are buffered by the presence of more than two sets of chromosomes.

The principle of balance is particularly well exemplified in man. Here most of the known chromosome mutations are associated with distinctive and abnormal phenotypes. The most common viable numerical variants are those involving the sex chromosomes (Table 1). These include a complex assortment of types with aneuploid states as high as pentasomy. Such sex chromosome aneuploidies are important causes of sterility, intersexuality and, occasionally, mental retardation. The occurrence of such extensive viable variation in sex chromosome make up stems from the fact that the Y-chromosomes of male somatic cells and all but one of the X-chromosomes in all somatic cells are heterochromatic, late replicating, and hence relatively inactive genetically.

By contrast with this situation only four viable autosomal trisomies are known in man. These involve autosomes 8, 13, 18 and 21. These chromosomes contain the highest proportion of Q- and G-banding material and, like the Y and the facultatively heterochromatic X, have relatively large regions that are late replicating. Because of this they are also presumed to be relatively inert genetically. Even so the presence of extra members in all of them leads to gross disturbances of phenotype which are generalised and major in form, producing states of mental deficiency. Since many thousands of children and adults have now been tested for chromosome mutations it seems valid to conclude that all other

Table 1 A summary of the known viable chromosome mutations in man.

		Chromosomes Involved	
	Type of Mutation	*Autosomes*	*Sex Chromosomes*
(1) Numerical	Monosomy (2x−1)	21	XO (Turner ♀)
	Polysomy	Primary trisomics (2x + 1)—8, 13, 18 and 21 (Down's Syndrome) Partial trisomy—short arm of 10, 17 and 20 Double trisomic (2x + 1 + 1)—18 + 21 (Down's Syndrome)	XXY, XYY, XXYY, XXXYY and XXXXY (Klinefelter ♂'s) XXX, XXXX, XXXXX (superfemales)
	(b) Mosaicism	Diploid/triploid (2x/3x)	XX/XY, XY/XXY, XX/XXY, XXXY/XXXXY, XX/XXX, XX/XXY/XXYYY XXXY/XXXXY/XXXXXY
(2) Structural	(a) Deletion	Part of short arm of 4 and 5 (Cri-du-chat Syndrome), 18_S, 18_L Short arm or satellites of 21 (Christchurch chromosome, Ch)	Probably X_L and Y_L
	(b) Ring-chromosomes	Various	X
	(c) Iso-chromosomes	Iso-21_L (Down's Syndrome)	Iso-X_L and iso-X_S (Turner ♀'s)
	(d) Pericentric inversion	1, 2, 3, 5, 8, 18 and 19	
	(e) Interchange (Reciprocal translocation)	1–9 2–10 4–6 5–11 8–15 11–12 1–11 2–15 4–9 5–16 8–18 11–13 1–13 4–10 11–17 1–14 3–13 4–13 6–12 9–16 11–19 1–15 3–19 4–17 6–13 9–22 1–16 3–20 4–18 6–14 19–22 1–17 1–18 7–11 1–19 7–19	X–2, X–14, X–18 X–Y in ⚥
	(f) Whole arm transposition	Common: 13–14, 14–21 Rare: 13–21, 13–22, 14–15, 14–22, 15 21, 13 [illegible] and 21–22 Homologous: 13–13, 14–14, 21–21	
	(g) Centric fusion	13–13	

autosomal aneuploids are probably lethal in the prenatal period. This is supported by the fact that whereas in new-born humans approximately 0.53% of a sample of 18 911 live-born children have been shown to have abnormal chromosomes, 42% of the 747 spontaneous abortions examined have been associated with detectable chromosome aberrations, the most common single class being that of autosomal trisomics (180 cases). Polyploidy is also relatively common; triploids, for example, constitute the second most common class of abortuses (60 cases).

(b) STRUCTURAL LIMITATIONS The ability of a chromosome to move on both mitotic and meiotic spindles depends on its capacity to establish reversible binding of the chromatid with the microtubule protein of the spindle system. A majority of plants and animals have only a single specialised site for spindle attachment. In such monocentric systems it is clear that this function cannot be simply acquired by other regions since both spontaneous and experimentally produced fragments lacking a centromere (acentric) fail to move in a regular manner. The same is usually true of dicentric chromosomes, that is chromosomes possessing two centromeres. In monocentric systems, therefore, the possibilities for karyotype change are limited by the fact that only chromosomes with single centromeres (monocentric), or in some cases fragmented centromeres, can survive.

Attitudes relating to the types of structural rearrangement that can contribute to chromosome variation in natural populations have been much influenced by the assumption that stable telocentric chromosomes do not occur. One consequence of this has been to argue that changes in chromosome number conditioned by prior structural changes have tended to be undirectional involving whole arm exchanges which have been erroneously referred to as centric fusions. There is now clear evidence that the centromere of at least some metacentrics is a compound element which can be divided by breakage into smaller, yet still functional, entities. More specifically some metacentrics can be subdivided into two equivalent telocentric entities. In this way one can account both for the production of new centromeres and new chromosomes. Thus although new centromeres cannot arise *de novo* they can be multiplied by the subdivision of existing ones.

There are still some authorities who argue that the telocentrics so produced are not stable and so do not form part of the normal karyotype of any species. In their view all chromosomes must be two-armed entities though they admit that the second arm may be so small as to be beyond the limits of consistent resolution with a light microscope. It is certainly true that some experimentally produced telocentrics are known to be unstable since they are subject to irregular disjunction leading to partial trisomy, to loss, or else are converted into iso-chromosomes, i.e. metacentrics with genetically identical arms which are mirror images of one another (p. 27). An equivalent instability is, however, known in many other induced chromosome changes (p. 20) so that this is not a singular feature of telocentrics. There are two recent lines of evidence which confirm the existence of stable telocentrics.

(i) Electron-microscope observations of the chromosomes of

Melanoplus and *Mus*, which are commonly referred to as acrocentrics, fail to show any short arms.

(ii) There are now several convincing cases involving the production of stable telocentrics in natural populations. *Nigella doerfleri*, for example, is a plant species which is endemic to Greece. Here the standard karyotype ($2n=12$) consists of five pairs of metacentrics and one pair of acrocentrics and is found in numerous populations taken from the Kikladhian islands. A deviating number, $2n=14$, has been found in some plants raised from seed collected on the island of Ios. These 14-chromosome plants have two pairs of truly telocentric chromosomes which give no idication of any instability in either mitosis or meiosis. These have arisen by centric fission of metacentric pair number 2 of the standard complement.

Reciprocal crosses between 12- and 14-chromosome plants lead to the production of 13-chromosome hybrids which are indistinguishable from the parent in morphology, pollen fertility and seed set. At meiosis five bivalents and one trivalent are regularly produced and by self-fertilizing such 13-chromosome hybrids the following products have been obtained:

Plant No.	*2n*			*Total*
	12	13	14	
1	13	22	16	51
2	12	22	6	40
3	11	21	18	50
Totals	36	65	40	141

A comparable case exists in the root vole, *Microtus oeconomicus*, which has an extended distribution in N. Europe and Asia. Here all animals from 5 different localities in Lapland, one locality in N.E. Finland, one in the northern Kuriles, one in DDR and one in Holland show $2n=30$ with all the members bi-armed. The related species, *M. kikuchii* and *M. montebelli* also both have $2n=30$ so there seems little doubt that this is the standard diploid karyotype in *oeconomicus*.

By contrast in Harjeddlen (Sweden) two out of six males caught in the wild had $2n=31$ and four had $2n=32$, while among offspring born in captivity seven had $2n=31$, seven had $2n=32$ and one only had $2n=30$. In animals with $2n=31$ one of the eighth chromosomes in order of size is replaced by two telocentrics while in the $2n=32$ form both metacentrics were so replaced. Here, as in *Nigella*, the fission is stable and transmissible.

(c) MEIOTIC LIMITATIONS Meiosis depends for its proper functioning on a certain degree of similarity between the sets of homologous chromosomes present in the chromosome complement. If homologous chromosomes do not pair or else do not segregate then aneuploid gametes will result. Aneuploid gametes lead to aneuploid zygotes and hence to zygotic death. Many chromosome mutations interfere with the linearity of pairing and this has the general effect of reducing gametic efficiency. To persist, either chromosome mutations must be able to pass through meiosis without producing aneuploid gametes or else some system must exist to compensate for their production.

In autopolyploids, for example, there are 4 homologous potential pairing units. This in itself produces no special problem for more than two homologues are perfectly capable of pairing and crossing-over provided they do so at different regions along the length of the paired units. However, autotetraploidy does give rise to an increased number of possible pairing patterns since any 4 homologues in an autotetraploid can form a quadrivalent, a trivalent plus a univalent, 2 bivalents, 1 bivalent and 2 univalents or else 4 univalents (see *Studies in Biology* no. 12) and most usually produce a mixture of these (Fig. 2–5). This, in turn, leads

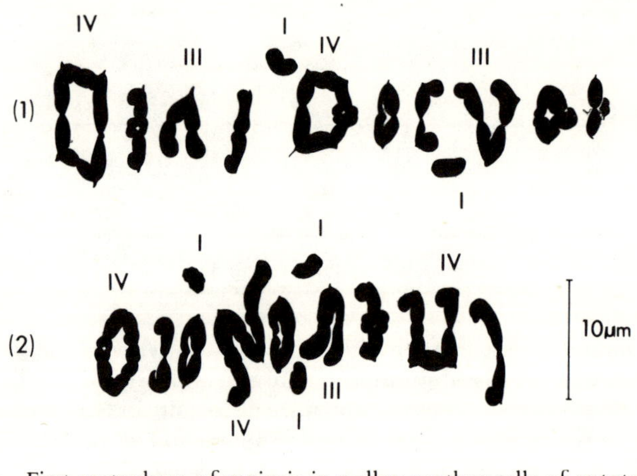

Fig. 2–5 First metaphase of meiosis in pollen mother cells of autotetraploid *Lolium perenne* ($2n=4x=28$, after Crowley and Rees). The upper cell (1) contains two ring quadrivalents, two trivalents and two univalents ($2n=2\text{RIV}+2\text{III}+2\text{I}+6\text{II}$). The lower cell (2) has one ring and two chain quadrivalents, one trivalent and three univalents ($2n=\text{RIV}+2\text{CIV}+1\text{III}+3\text{I}+5\text{II}$).

to problems in segregation for only gametes carrying one or more complete haploid sets will be balanced.

It is of course possible to regularize meiosis by forming ring quadrivalents or bivalents only, or else mixtures of these, for both these types tend to give numerically equal segregation. However, should even this occur a further problem may arise at fertilization. If the autotetraploid is crossed with its diploid parent it gives rise to an autotriploid. Autotriploids are almost always sterile for their segregational problem is an extreme one. Consequently little can be done to improve sexual fertility in such an uneven polyploid and triploids are compelled to adopt either an asexual or else an apomictic (*plants*) or parthenogenetic (*animals*) mode of reproduction in order to survive (p. 24).

The meiotic problems posed by chromosome mutation are no less acute in structural rearrangements. Like all other forms of mutation, such rearrangements are relatively rare events so that the possibility of both homologues in a diploid being affected in the same way at the same time is remote. Consequently all structural rearrangements are introduced into natural populations in a chromosomally heterozygous state. If two such structural heterozygotes cross with one another they will of course be expected to produce structural homozygotes, in which both homologues are affected, in one quarter of their progeny. One needs therefore to distinguish three karyomorphic states for any type of exchange—basic homozygotes, in which both homologues are unaffected; structural heterozygotes, in which one homologue is rearranged and structural homozygotes, in which both members of an homologous pair are rearranged. Structural homozygotes, like basic homozygotes, give rise only to bivalents at meiosis but structural heterozygotes lead to unusual pairing configurations. Thus, in paracentric inversion heterozygotes homologous pairing is possible only if a reverse pairing loop forms (see *Studies in Biology* no. 21). Where such reverse loops do form and single chiasma occur within them the products of crossing-over are duplicate/deficient for certain segments and this leads to aneuploid gametes. Likewise in interchange heterozygotes gametic unbalance arises either as a result of non-disjunctional arrangements of the various meiotic associations at first anaphase (Fig. 2–6) or else as a result of crossing-over within the segments between the centromere and points of exchange (the interstitial segments). Three main factors influence the degree of disjunction in this case:

(i) the frequency and distribution of chiasmata—disjunction frequency tends to be highest where the chiasma frequency is minimal compatible with consistent multiple formation and where chiasmata are terminal in location and so do not occur within the interstitial segments;
(ii) the symmetry of interchange—disjunction frequency tends to be highest where the interchanged segments are approximately of equal length and where, therefore, the interchange is symmetrical. This leads to an equidistant distribution of centromeres in the mutliple and so facilitates its flexibility;

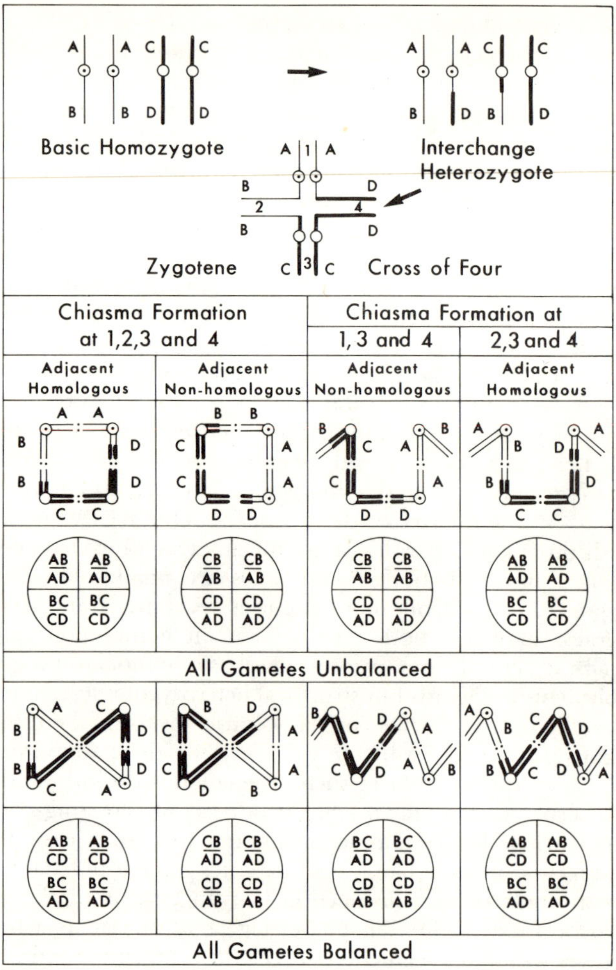

Fig. 2–6 The meiotic consequences of structural heterozygosity for a single interchange following the formation of maximum ring or chain of four multiples. Note only alternate orientation of homologous centromeres (lower part of figure) produces balanced genetic combination in all four meiotic products.

(iii) the number of chromosomes involved in the interchange—in general the larger the interchange, and hence the maximum multiple which may form, the lower is the expectation for disjunction.

It is unlikely, therefore, that a newly produced interchange will become incorporated into a polymorphic system within a species unless a high frequency of disjunction is present at the time of its inception.

It has often been argued that heterozygosity for a whole arm transposition or a centric fusion/fission is unlikely to lead to a reduction in fertility because one might, as a general rule, expect an orderly segregation of two acrocentrics (transposition) or two telocentrics (fusion/fission) from the single metacentric present in the heterozygote. While this expectation is fulfilled in some trivalents (see p. 28) it is clearly not applicable to others. The grey house mouse (*Mus musculus*) has 40 telocentric chromosomes. The tobacco mouse (*Mus poschiavinus*), which is closely related to it, has a diploid count of 26 and is homozygous for seven fusions. In F_1 hybrids between them seven trivalents are regularly formed. By studying second metaphase counts it can be shown that 50% or more of the secondary spermatocytes in these hybrids are unbalanced, a consequence of non-disjunction at first anaphase (Fig. 2–7). By studying

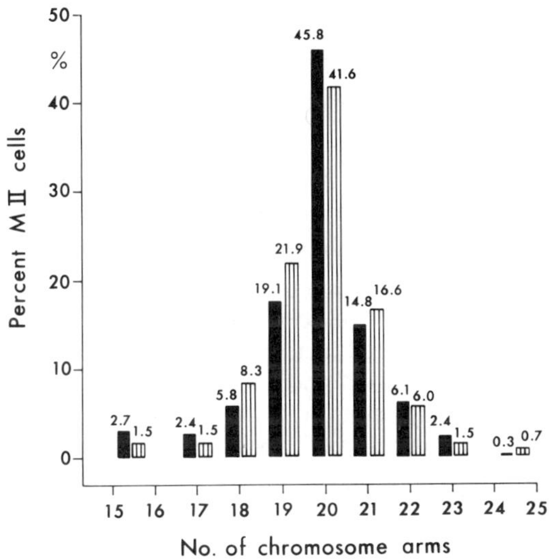

Fig. 2–7 Frequency distribution histograms of the number of chromosome arms present in second metaphase cells obtained from two male F_1 ($2n=33=7$ metacentrics + 26 telocentrics) hybrids (shown solid and hatched respectively) between *Mus musculus* ($2n=40$ telocentrics) and *Mus poschiavinus* ($2n=26=14$ metacentrics + 12 telocentrics). Balanced metaphase-II cells are those containing 20 chromosome arms. (From TETTENBORN and GROPP, 1970, by permission of S. Karger AG, Basel.)

hybrid derivatives carrying only one of the seven metacentrics present in the F_1 there is clear evidence for striking differences between animals heterozygous for different metacentrics in respect of the non-disjunction frequency (Table 2). Whether these differences depend largely on variations in chromosome size and chiasma number and position, and their subsequent effects on trivalent orientation, is not known but

Table 2 Disjunction behaviour and fertility effects of individual metacentric chromosomes of *Mus poschiavinus* introduced by crossing into a predominantly *Mus musculus* background. Note that most of the trisomic embryos die before birth; the few that survive to this stage die shortly afterwards (data of FORD AND EVANS, 1974).

	Percentage non-disjunction			*Estimated reduction in fertility*
Metacentric	*Estimated from MII-data*	*Estimated from trisomic implants*	χ^2	*%*
T_1 Bnr	19.6	22.3	0.03, $P>0.8$	21
T_2 Bnr	33.5	25.0	1.53, $P<0.2$	29
T_3 Bnr	16.1	9.1	0.64, $P<0.3$	13
T_7 Bnr	10.7	5.7	2.77, $P\sim0.1$	9

considerable heterogeneity can be shown to exist between individuals heterozygous for the same metacentric.

(d) UNDEFINED FACTORS In both spontaneous and induced translocation heterozygotes of mammals infertility or subfertility appears to arise in two quite distinct ways:

(i) by the production of aneuploid gametes which, though they function at fertilization, lead to pre-implantation zygotic death or to abortion of the unbalanced foetus;
(ii) by gametogenic arrest, a phenomenon which is either confined to, or at least predominates in, male heterozygotes where it leads to the production of few (oligospermia) or no sperm (azospermia).

As far as spermatogenic arrest is concerned the time of breakdown appears to be variable. In mice a majority of X-autosome translocations, T(X:A), lead to meiotic breakdown at pachytene. The same is true of at least some T(Y:A) males. In human A:A translocations breakdown starts between first and second division and ends in complete arrest during spermiogenesis (see also p. 62).

On *a priori* grounds there is no reason why translocation heterozygosity should lead to spermatogenic failure. Several authors have suggested that such gametogenic disturbances are the result of position effects, that is disturbances of gene activity associated with the translocation. As yet, however, no one has provided convincing evidence of this. Spermatocyte degeneration has also been reported in a trisomic mouse, the sterile son of a male treated with a chemical mutagen. Unless this was a tertiary trisomic, produced by abnormal segregation from a stem cell already heterozygous for a translocation, which seems unlikely, sterility cannot sensibly be attributed to a position effect in this instance.

2.3 Compensation mechanisms

At least two mechanisms are known which compensate for the formation of unbalanced chromatids and so avoid the production of aneuploid gametes. Both operate in paracentric inversion heterozygotes. Here crossing-over within the limits of the inversion leads to the production of complementary duplication/deficient dicentric and acentric cross-over chromatids. In several female Diptera, including *Sciara* and *Drosophila*, where such heterozygotes are commonly found in nature (p. 51), the chromatid bridges produced by the dicentric at first anaphase persist into the second meiotic division. The effect of such persistent dicentric ties is to ensure that one of the balanced non cross-over chromatids passes preferentially to the innermost nucleus of the essentially linear tetrad. It is this nucleus that develops into the functional female nucleus, the remaining three meiotic products transform into polar nuclei which degenerate. Thus there is no loss of gametes and all the gametes produced are balanced (Fig. 2–8).

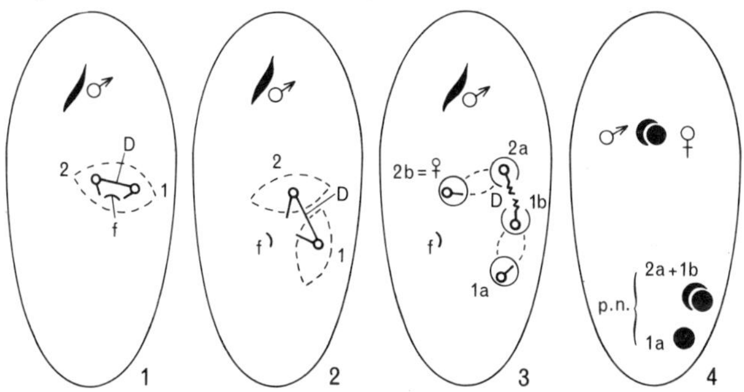

Fig. 2–8 Diagrammatic representation of the cytological behaviour of a telo-dicentric bridge (D) and an acentric fragment (f) produced following crossing-over in a paracentric inversion in *Sciara impatiens*. The dicentric stalls at first anaphase (1) leading to a preferential orientation of the two centromeres of the dicentric at second division (2) and the preferential movement of non-cross over chromatids (3) to the terminal nuclei (2b and 1a) of the meiotic tetrad. One of these nuclei (2b) becomes the definitive egg nucleus (4) the other three forming polar nuclei. (After Carson.)

Equivalent persistent bridge ties are known also following crossing-over within paracentric inversion heterozygotes in male chironomids and blackflies. Here they result in the production of giant spermatids which are non-functional. While this results in a reduction in the number of functional sperm such a reduction is unlikely to be of significance because of the vast number of sperm produced by an individual. It does, however, ensure that no balanced eggs are sterilized by unbalanced sperm. This problem does not exist in *Sciara* and *Drosophila* because male meiosis is achiasmate in both cases.

3 Chromosome Variation in Natural Populations: I. Systems of Variation

Facts are building stones of Science but Science does not begin with facts. It begins with observations.

C. J. Herrick

3.1 Numerical variation

3.1.1 Polyploid complexes

One of the most striking differences in the cytogenetic structure of plant and animal populations relates to the fact that sexual dioecism, which has become a well established outbreeding mechanism in animals, is rare among plants where alternative mechanisms have been used to secure outbreeding. It was Muller who first linked this difference with the widespread occurrence of polyploidy in plants and its relative rarity in animals. More specifically he argued that, in organisms with well defined sex-chromosomes, polyploidy will produce inter-sexuality and sterility. Apparent support for this conclusion comes from the fact that polyploid species are best known among animals in fish and amphibians, two groups where differentiated sex chromosomes are rarely if ever found. Muller's argument is also supported by evidence from *Drosophila* and *Rumex acetosa.* Here the X-chromosome is female-determining and the autosomes male-determining. Tetraploids are therefore XXXX (♀) and XXYY (♂) in constitution and crosses between such individuals give a proportion of XXY types which are sterile intersexes. Even so it is now clear that Muller's case has been overstated since such a sterility barrier is not established in cases where the Y-chromosome is strongly male-determining, as in the plants *Melandrium, Acnida* and *Rumex acetosella* where tetraploid XXXY individuals are males. Here, therefore, crosses between XXXX females and XXXY males give constant tetraploid and dioecious strains and intersexuality is exceptional.

Polyploidy is one of the most common changes found within and between species of Angiosperms. The primary adaptive advantage of increasing the number of genomes in this way is in genetic buffering or stabilization which allows for a greater exploitation of the interaction between genomes. The most significant forms of polyploid evolution in both plants and animals are unquestionably those which have evolved in combination with hybridization. Here polyploidy serves to stabilise useful hybrid products, partly by eliminating the sterility characteristic of hybrids and partly by reducing the amount of genetic segregation in the hybrid. There are, however, three rather distinct types of polyploid complexes:

(a) INTER-ECOTYPIC HYBRIDS The races of Cocksfoot, *Dactylis glomerata*, found north of the Mediterranean region in Eurasia are tetraploids. Morphologically and ecologically these races are intermediate between two diploids. One of these, confined to the forests in central and north Europe, is known as *D. Aschersoniana*; the other, which inhabits the semi-arid steppe country of S.W. Asia, is known as *D. Woronowii*. Artificial hybrids between these diploids are fully viable and fertile both in the F_1 and F_2 generations so that although they are given specific status they are in fact different ecotypes of the same species. In a like manner many other tetraploid subspecies of this genus in S. Europe, N. Africa and W. Asia combine together the characteristics of different diploid subspecies.

(b) INTROGRESSIVE HYBRIDS Backcross derivatives of tetraploid and diploid populations are sometimes superior to either of the parents in a particular ecological niche because of the unilateral introgression of genes from the diploid to the tetraploid. Such gene flow increases both the morphological range of variation and the ecological range of tolerance of many tetraploids. Alternatively the gene pool of a tetraploid can be enlarged and expanded by secondary hybridization and introgression with related tetraploids. This process is particularly effective when the hybridizing tetraploids share one common pivotal genome. For example in the grass genus *Aegilops, A. umbellata* forms a pivotal diploid species from which seven distinct tetraploids, all forming aggressive weeds which are widespread in the Mediterranean region, have been derived. These combine one genome of *umbellata* with a second genome derived from one of three different diploid species complexes —*A. caudata, A. corrosa* and *A. speltoides*. In these mixed populations, hybrids are often found between two different tetraploids. Such hybrids are pollen sterile but set a small number of seeds from open pollination by the parental species. The progeny so obtained recover fertility in one or two generations and these fertile introgressed genotypes become fixed by self-pollination. Nevertheless they remain inter-fertile with other individuals of the species concerned because selfing is never complete.

(c) APOMICTIC AND PARTHENOGENETIC POLYPLOIDS In sexual reproduction there is a regular alternation between meiosis and fertilization. Systems have, however, arisen in which either meiosis or fertilization, or else both these processes, have been dispensed with. In plants such systems are grouped under the collective term apomixis. In animals they are known as parthenogenetic systems. Several plant polyploid complexes reproduce apomictically. Thus most of the races of the dandelion, *Taraxacum officinale*, found in N. and W. Europe are triploid and obligate apomicts.

Comparable cases are known in animals. For example in the whiptail lizard, *Cnemidophorus*, interspecific hybridization and parthenogenesis have combined to produce triploid and tetraploid forms adapted for the utilization of weedy habitats. Triploid species have evolved as a consequence of hybridization between diploid parthenogenetic and diploid bisexual forms while hybridization between a triploid all-female parthenogenetic species, *C. sonorae*, and a diploid bisexual species, *C. tigris*, has resulted in the production of tetraploid forms.

A somewhat similar case occurs in the Arctic blackfly *Gymnopais* studied by Rothfels. Here two sister species are known—*holopticus* and *dichopticus*—which are distinguished by four inversions distributed over the three chromosomes which make up the haploid set ($2n=2x=6$). Two kinds of parthenogenetic triploids are also found combining two sets of chromosomes from the female of one species with one set from the male of the other. That is, one is 2H + 1D and the other 2D + 1H in constitution. Some populations contain all four kinds of individuals, others are pure tiploid.

What applies to allopolyploids holds also for autopolyploids. Thus diploid sexual races and polyploid parthenogenetic races are well known in several invertebrate groups including flatworms, oligochaetes, crustaceans and insects.

3.1.2 Aneuploid systems

Naturally occurring aneuploid systems are rare in diploids, in keeping with the principle of balance. Thus even when aneuploidy is tolerated developmentally, as in the sex-chromosome aneuploids of man, it leads to sterility. It is true that naturally occurring plants with 13, instead of the usual 14, somatic chromosomes have been found in several populations of *Clarkia amoena* subspecies *huntiana* in N. America. These plants show no obvious phenotypic differences from their 14-chromosome sibs. However, although referred to as monosomics, analysis of meiotic pairing within them indicates that they are products of a complex system of reciprocal translocations involving three non-homologous members. This is coupled with the loss of a small chromosome consisting of a centromere and the heterochromatic regions which flank it. They are therefore in reality pseudo-monosomics.

By contrast with the situation found in diploids, aneuploid derivatives are not uncommon in polyploids. *Ophiopogon japonicus* and *O. ohwii*, for example, are perennial liliaceous species which grow commonly in fields and forests of E. Asia. In *O. japonicus* both diploids and tetraploids are known ($2n=2x=36$ and $2n=4x=72$) while *ohwii* is essentially a tetraploid ($2n=4x=72$). Dimorphic hypotetraploids are known in both species with $2n=4x-5=67$ and $2n=4x-4=68$ respectively. Moreover in material collected from Tokyo and Okinawa the standard forms are not found at all and all specimens are hypotetraploid. The $4x-5$ form is quite viable despite its apparently unbalanced genome. Indeed it shows an extensive distribution, a relatively high frequency of good pollen and a predominance of mother-type plants among its progeny. It appears therefore that a balanced change has been established by genetic restructuring at the aneupolyploid level. A somewhat similar case occurs in the spread of the highly successful *Spartina townsendii* complex in Britain. The present day form is an aneuploid backcross derivative of an original allopolyploid produced between the hexaploid *S. maritima*, a species native to Britain, and *S. alterniflora*, an octosomic hexaploid introduced into Southampton Water through shipping ballast (Fig. 3–1).

Perhaps the most extensive series of chromosome numbers found in any group of closely inter-related populations is that which occurs in the genus *Claytonia*, the spring beauties of eastern N. America. For example *Claytonia virginica* (Portulaceae) is thought to have evolved by hyperaneuploidy from a base of $n=6$ to $n=x=7$ and 8, with each race giving rise to widespread and dominant primary tetraploids ($n=2x=12$, 14 and 16). These, in turn, have formed many secondary races largely by hypoaneuploidy ($n=2x-1=11$ and 15, derived respectfully from the $n=2x=12$ and 16 types) as well as higher polyploids ranging from $6x$ to $12x$ in cases where $x=6$, and $6x$ and $8x$ where $x=7$. In spite of this

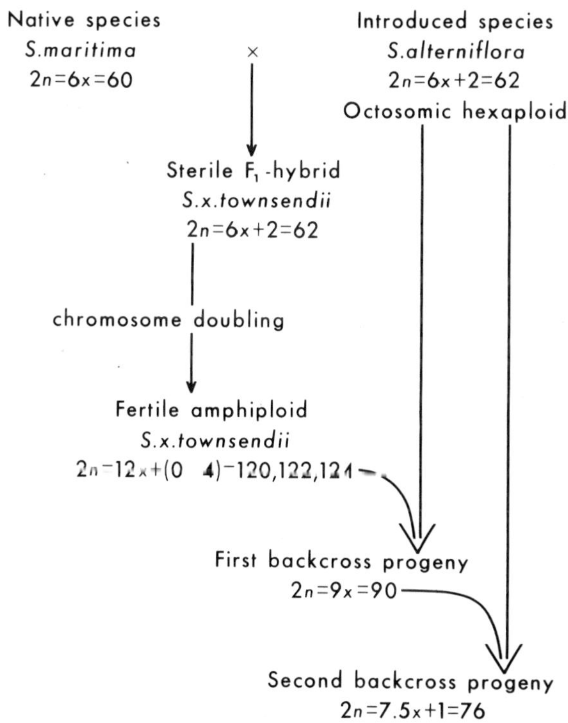

Fig. 3–1 Origin of the chromosome races of the *Spartina townsendii* complex in Britain. (After Marchant.)

extensive chromosomal variation, the overall range of morphology and ecology within the genus is narrow and the various cytotypes appear to represent variations on a restricted theme.

3.1.3 Supernumerary chromosome systems

Despite the infrequent occurrence of aneuploidy in diploid systems many diploid species are now known to carry extra or supernumerary chromosomes often referred to as B-chromosomes. These constitute a heterogeneous class which at one extreme grade into the standard or A-chromosomes in terms of structure and behaviour but most usually are clearly distinguishable from, and lack homology with, members of the standard set. These chromosomes range in size from the smallest to the largest elements in the species (Figs. 3–2e and 3–3a) and they are often, though not always, heterochromatic in character. Where B-chromosomes are present their number and frequency varies from population to population. The highest recorded numbers are in plants and in some populations of *Taima laxifolia* from S.W. Japan, B-chromosomes have been found in every individual examined.

In only one case is the origin of such supernumeraries known with any certainty. This is in *Chironomus plumosus* and *Ch. melanotus* where their structure and meiotic behaviour indicates that they are centromeric fragments of the smallest member of the complement. Here, presumably, the B must have originated by non-disjunction followed or accompanied by loss. A comparable origin can be inferred for many other kinds of

---

Fig. 3–2 First metaphase of male meiosis to illustrate the various kinds of chromosome mutations. (**a**) *Chorthippus brunneus*. A chain of four multiple (arrow) showing alternate (disjunctional) orientation in an individual heterozygous for a single reciprocal translocation ($2n = 1IV + 6II + X$). Normally in this species there are eight bivalents, three of which are long metacentrics (compare Fig. 2–2a). In this photograph there are only two metacentric bivalents; the third is involved in the interchange multiple together with one of the telocentric pairs. (**b**) *Secale cereale*. Rye grass normally has seven bivalents. In this mutant there is an interchange chain of six multiple in non-disjunctional orientation ($2n = 1VI + 4II$). (**c**) *Cryptobothrus chrysophorus*. This individual is heterozygous for two centric transpositions (two left arrows) and homozygous for a third (top right arrow). It is also heterozygous for a supernumerary segment in the smallest pair of autosomes (bottom right arrow and compare Fig. 2–2e and f). (**d**) *Oedaleonotus enigma*. This species is characterized by an XY male, XX female sex chromosome mechanism and the XY bivalent is located at the top of the photograph. The individual shown is also heterozygous for a fusion between two medium-sized non-homologous autosomes and the fusion trivalent (arrow) is oriented to give disjunctional separation of its three members. (**e**) *Buforania crassa*. There are five telocentric mini B-chromosomes (arrows) in addition to the conventional twenty three telocentric members. (**f**) *Secale cereale*. A tetraploid cell of rye with two ring-quadrivalents (arrows) in addition to ten bivalents ($2n = 4x = 28 = 2IV + 10II$).

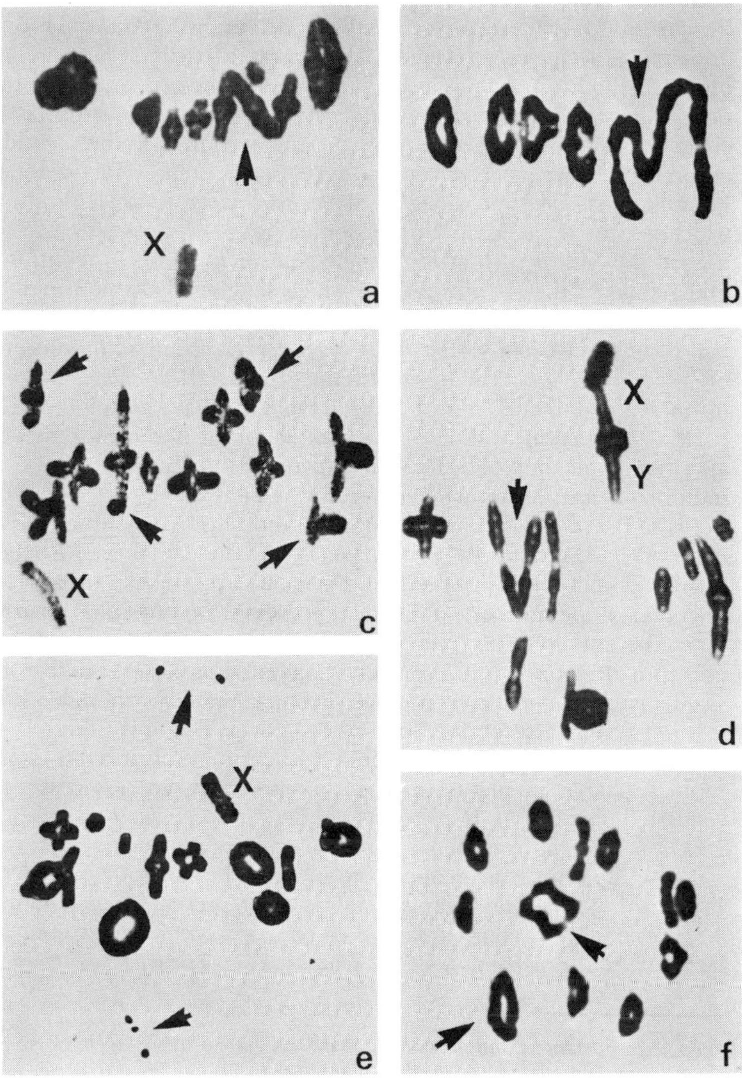
X
a
b
X
c
X
Y
d
X
e
f

B-chromosomes. Thus some supernumeraries are iso-chromosomes, or iso-chromosome derivatives, implying an origin by misdivision of unpaired telocentrics, an event which is known to occur not uncommonly in univalent chromosomes at first anaphase of meiosis. In insects it has been widely believed that B's are derived from X-chromosomes. This claim is, however, based largely on the fact that both X- and B-chromosomes are heterochromatic in character. Better evidence for X–B homology has been provided in *Myrmeleotettix maculatus*. Here the B^m, which is a metacentric iso-chromosome (Fig. 2–1d) possesses a G-band toward the end of each arm. A single equivalent band is present in the mid-region of the X-chromosome which suggests that the B may have originated by misdivision of the X together with the deletion of the distal half of it. In fact many B's appear to undergo secondary modification, often involving loss. The most extreme case is *Aster ageratoides* where 25 distinct though related morphological types of B have been found.

Non-disjunction leading to the production of B-chromosomes may also sometimes occur following unequal distribution from chain multiples in translocation heterozygotes, as has been suggested by Harlan Lewis in the plant *Clarkia*. Here translocation heterozygotes give rise to trisomics with a relatively high frequency and these, in turn, are believed to give rise to B-chromosomes. The *Clarkia* B's are indistinguishable from members of the normal complement at metaphase but do not pair with them, or usually with one another, despite the fact that they are euchromatic. During mitotic prophase they do not show the diffuse distal segments characteristic of normal chromosome arms though whether this is a result of secondary loss is not known. The only other possible mode of origin of supernumeraries is by hybridization and this is clearly more probable in plants than in animals. Such an origin has been suggested for B's in *Haplopappus gracilis, Anthoxanthum odoratum* and *Oenothera* though the evidence is not particularly convincing.

Many supernumeraries appear to be devoid of any obvious effect on the morphology of the organism, at least when present in small numbers. It is presumably this that explains their common occurrence in contrast to conventional polysomics. It is true that in *Haplopappus gracilis* B-

Fig. 3–3 Further examples of first metaphase of male meiosis showing chromosome mutations. (**a**) *Phaulacridium vittatum.* This individual carries a single large supernumerary chromosome in addition to the eleven autosomal bivalents and the univalent X ($2n = 23 + 1B$). (**b**) *Buforania crassa.* A ring of four interchange multiple (arrow) involving all telocentric elements (compare Fig. 3–2e). (**c**) *Percassa rugifrons.* A species with a diploid count of twenty one in the male (XO) and characterised by the presence of one large metacentric pair (arrow). Additionally this individual is heterozygous for a supernumerary segment on the smallest telocentric autosome (arrow). (**d**) *Secale cereale.* This cell is from a trisomic plant ($2n = 2x + 1$) and the three homologues have paired as a chain of three.

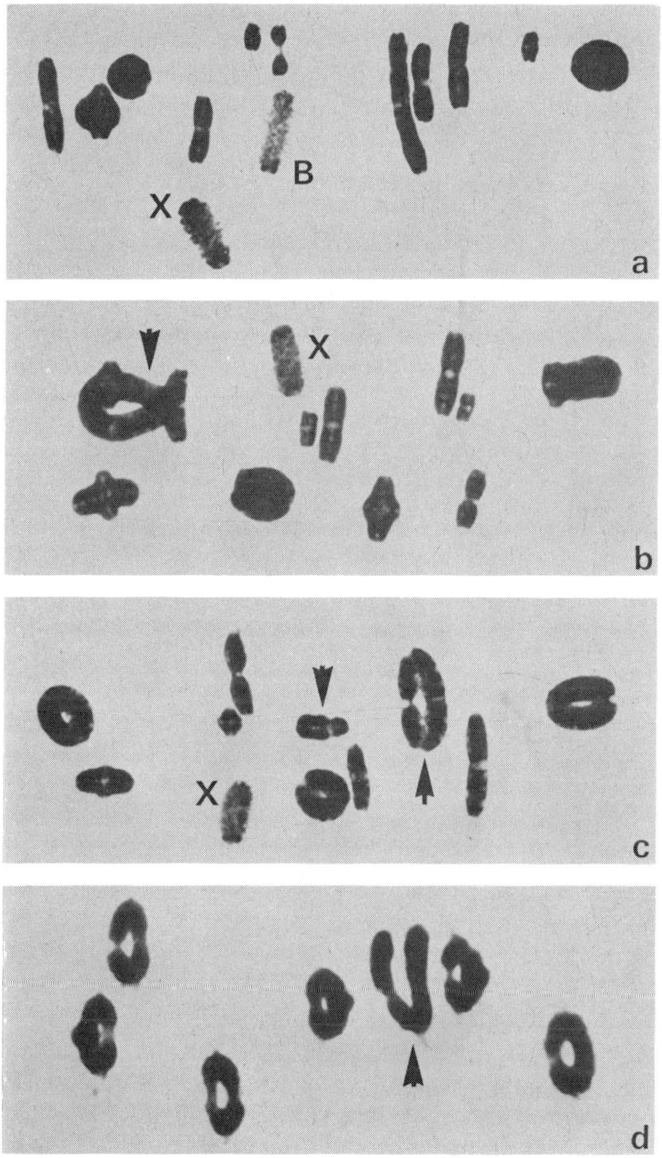

UNIVERSITY
OF BRISTOL
LIBRARY
VETERINARY
SCIENCE

chromosomes are known to produce morphological changes in leaves, stems and achenes. This species is, however, unusual in its exceptionally low number, $2n=4$, so that such effects are perhaps not surprising. Likewise in plants where particularly high numbers of B's are present, and sometimes even where the B-number is low, it is possible to demonstrate clear affects of B's on vegetative vigour, on endosperm development and on fertility. This applies, for example, to *Anthoxanthum aristatum, Secale cereale, Festuca pratensis* and *Lilium callosum*. Such effects, of course, are commonly found in conventional trisomics.

A majority of the supernumerary systems known in nature are characterized by boosting mechanisms which lead to increases in B-frequency in the progeny of B-containing individuals. Five rather different systems have been described.

(a) In *Locusta migratoria* while B-chromosomes show no variation in number within the cells of the gastric caeca, the only somatic tissue from which diploid mitoses can be obtained in the adult, they may vary in number in the primary spermatocytes of the same individuals (Table 3).

Table 3 The number of B's in the cells of the gastric caeca and in the primary spermatocytes of 10 males of *Locusta migratoria* (data of KAYANO, 1971).

Male	*Gastric caeca*	0B	1B	2B	3B	4B	5B	6B	*Mean no. B's per cell*	*No. cells observed*
				% Primary spermatocytes with						
No. 22		1.5	96.2	2.3	—	—	—	—	1.01	1346
No. 13		1.8	78.3	19.8	0.1				1.18	1384
No. 28	1B	6.0	72.0	20.0	2.0	—	—	—	1.18	1413
No. 9		2.4	79.2	12.2	6.2	—	—	—	1.22	1855
No. 8		5.1	68.6	24.3	—	2.0	—	—	1.25	1561
No. 27		0.6	61.1	30.3	4.0	4.0	—	—	1.50	1450
No. 2		—	8.0	78.5	9.3	4.1	0.1	—	2.10	1385
No. 4	2B	—	2.1	86.5	8.3	3.1	—	—	2.12	1772
No. 6		—	8.0	36.6	34.7	18.3	2.4	—	2.70	924
No. 24		—	12.4	27.1	29.9	23.5	6.8	0.3	2.86	1327
Means		1.74	48.59	33.76	9.45	5.50	0.93	0.03	1.712	—

The pattern of variation indicates that the B's must accumulate in the germ line. The rate of accumulation varies from 1–50% in males with 1B and from 5–43% in males with 2B's. This pattern of distribution appears to be determined primarily by non-disjunction of B's in the mitoses associated with the initial differentiation of the testis follicles and is preferential in that cells with increased B-number are more frequently present than those with decreased B-numbers. Equivalent germ line instability is known also in the grasshoppers *Camnula pellucida* and *Calliptamus barbarus*. Such an increase of B's within the germ line following preferential non-disjunction in the pre-meiotic mitoses clearly increases the likelihood of transmission of the B to progeny.

(b) Preferential segregation of the B-chromosome at female meiosis is known in both animals and plants. Thus by comparing the B-chromosome constitution of progeny from single pair matings between parents of known constitution it has been shown that there is a preferential transmission of supernumeraries through female meiosis in several species of grasshopper (Table 4). This presumably results from the directed orientation of univalent B's to the definitive egg nucleus. Comparable behaviour is known also in the embryo mother cells of several plants including *Lilium callosum* and *Trillium grandiflorum*.

Table 4 Transmission frequencies (k̄) per B-chromosome in laboratory crosses of three species of grasshopper (data of LUCOV and NUR, 1973—*Melanoplus*; Hewitt—*Myrmeleotettix* and Rothfels (unpublished)—*Chloealtis* respectively). In the case of *Chloealtis* there is evidence of bimodality in k̄ values obtained from 1B♀ × 0B♂ crosses; that is some females accumulate and others do not.

	k̄ *per* B					
Species	0B♀ × 1B♂	1B♀ × 0B♂	1B♀ × 1B♂	0B♀ × 2B♂	2B♀ × 0B♂	2B♀ × 1B♂
Melanoplus femur rubrum	0.525	0.817				
	1.342					
Myrmeleotettix maculatus (a) East Anglia	0.38	0.58		0.26	0.67	
	0.96			0.93		
(b) Tal-y-Bont	0.49	0.89		0.41	—	
	1.380			—		
Chloealtis conspersa	0.38	0.66	0.63	0.32	0.79	0.68
	1.04			1.11		

(c) In Angiosperms the haploid product of meiosis (the gametophyte) undergoes two post-meiotic mitoses and directed non-disjunction of B's is known to occur in the first of these divisions in some species. In male gametophytes, for example, this division produces a vegetative and a generative nucleus. The former does not divide again but the latter does to produce two male gametic nuclei. In many grasses, including *Secale, Briza, Anthoxanthum* and *Dactylis*, B-chromosomes undergo polarized non-disjunction at the first division, passing preferentially to the generative nucleus. An equivalent behaviour is known also in the female gametophyte of rye.

(d) In *Zea mays* B's undergo non-disjunction at the second mitosis of the male gametophyte. This leads to a relative increase in the proportion of gametes with even numbers of B's or with zero B's. Coupled with this there is a preferential fertilization of the egg by B-carrying gametes. This accumulation mechanism, which selects sperm carrying B-chromosomes for preferential fertilization, breaks down, however, when several B's are present, which explains why large numbers of B's do not accumulate in natural populations.

(e) Distinctive boosting systems are known in bugs (Hemiptera), organisms that show non-localized centric activity and have an unusual meiotic behaviour. In animal bugs (Hemiptera. Heteroptera) the situation is best known in the bed bug, *Cimex lectularis* (♂ $2n=26+X_1X_2Y$; ♀ $2n=26+X_1X_1X_2X_2$). Here the sex chromosomes in the male do not pair in the first meiotic division but divide equationally. They pair and segregate at second division and then do so by a mechanism which does not involve crossing over. Populations from the New World and from various parts of the Pacific have no B's whereas populations from the Mediterranean region and elsewhere in the Old World have 4 B's which behave like extra X-chromosomes in the sense that they do not pair at first division and they always separate from the Y at second division of meiosis. This means, of course, that they accumulate since they are not subject to any reduction in number at male meiosis. It also means that they pass exclusively into the X-bearing gamete which, on fertilization, becomes a female.

In plant bugs (Hemiptera: Homoptera) too there is evidence for non-reduction of supernumeraries in male meiosis. Thus in *Pseudococcus citri* homologues do not pair in the first meiotic division but divide equationally. In the second division paternal and maternal homologues segregate from one another without pairing or recombination. Spermatids containing paternal homologues then degenerate. B-chromosomes segregate with maternal homologues at second division and pass, without reduction, into the functional gametes.

The existence of these boosting mechanisms has led to the widespread adoption of the hypothesis, first proposed by Östergren, that B-chromosomes have no useful function in the populations that carry them but are maintained in them by virtue of the efficiency of their accumulation systems. This suggestion has been reinforced by the demonstration that in some organisms B's appear to reduce both viability and fertility. In rye, for example, they cause a delay in germination and flowering and a general reduction in plant vigour and fertility. Such a reduction can be expected to lead to the loss of B's from a population and, for this reason, numerical increase by accumulation has been suggested as a balancing mechanism which maintains B-chromosomes within populations at equilibrium frequencies.

It is, however, characteristic of many B-chromosome systems that their frequency varies between populations within a species, often over wide ranges. Coupled with this there are good reasons, in at least some of these cases, for arguing that this variation is adaptive and can be related to the differing environments in which the populations exist. Thus surveys in the plants *Secale* (rye), *Phleum, Festuca, Centaurea, Dactylis* and in the grasshopper *Myrmeleotettix maculatus* indicate that the frequency of B-chromosomes is influenced by soil type and/or climate. In populations of rye from Korea the frequency of B's is higher in acidic soils of basic character. Again extensive sampling of natural populations of *Phleum phleoides* on the island of Oland in the Baltic has shown that the overall B-frequency in populations growing in sandy and moraine soils in N. and Central Oland is very much higher than in those growing on the so-called alvar areas, in the southern part of the island, comprising exposed or barely covered bedrock.

A metacentric iso-B-chromosome (B^m) is present in many southern British populations of the grasshopper *Myrmeleotettix maculatus* (Figs. 2–1 and 3–4). Even so, populations differ markedly in their B-chromosome content over relatively short distances and, additionally, populations in the south-west of the country carry a morphologically distinguishable iso-chromosome derivative which is submetacentric (B^{sm}). Two areas of this general distribution have been studied in some detail (Fig. 3–5).

(a) In Aberystwyth, West Wales, a cline of decreasing B^m-frequency can be demonstrated as one moves from the warm, dry coastal lowlands to the colder, wetter, mountains. Indeed in the higher mountains the species itself ceases to be represented.

(b) In East Anglia the coastal populations in the N. and E. of the region are devoid of supernumeraries whereas the inland populations in the S. and W. carry some 40–50% B^m-chromosomes. All the populations in E. Anglia occupy low elevation but the inland populations enjoy the highest summer temperatures.

Thus while no single climatic factor appears to be responsible for B-chromosome distribution in this species the general level of environmental stringency is important in maintaining the differential B-chromosome content of the various populations.

Direct evidence that rye plants with and without B's can show varying properties in respect of survival has been shown by B-behaviour in experimental populations. Sixteen stock populations with from 0–6 B's were established from a cross between *S. cereale* and *S. vavilovii*. These were planted in two replicates, each of 50 seedlings, at each of 2 sites at 2 densities and in 2 different seasons. The sites, Penglais and Syfydrin in Aberystwyth, Wales were at 17m and 270m respectively. These sites differ both in climate and in soil type. At Penglais the soil is well drained, well limed and fertile, whereas at Syfydrin it is poorly drained, acid and of low fertility. The mean B frequency was significantly higher among survivors at Syfydrin in comparison with Penglais and significantly lower

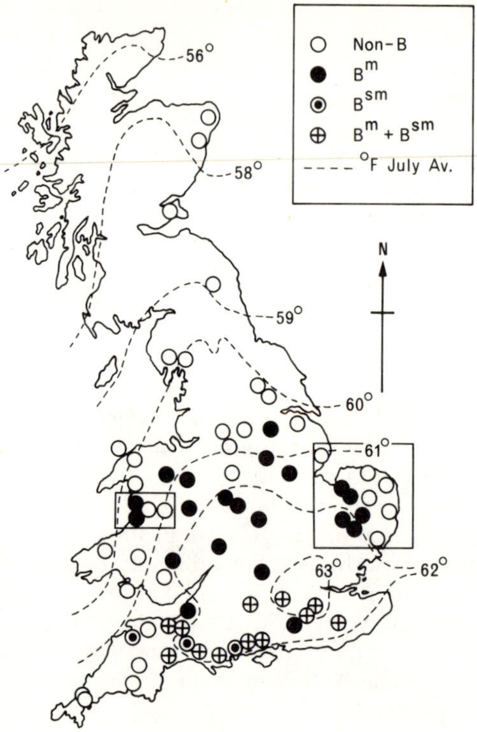

Fig. 3–4 The distribution of B-chromosomes in British populations of the mottled grasshopper *Myrmeleotettix maculatus*. The two enclosed areas have been subject to detailed analysis and the results of this are shown in Fig. 3–5. (From HEWITT and JOHN, 1967.)

among survivors in close populations (plant 12–15 cm apart in rows with 12–15 cm between rows) relative to open populations (plants sown 60 cm apart in rows with 60 cm between rows). Thus within one generation selection, operating through differential mortality, has resulted in a marked alteration in B frequency (Fig. 3–6).

In some few species there is no known mechanism for numerical increase in B-chromosomes. This includes *Centaurea scabiosa, Anthurium*

Fig. 3–5 The distribution of B^m-chromosomes in populations of *Myrmeleotettix maculatus* from East Anglia (above) and Aberystwyth (below) in relation to the mean daily maximum June temperature (East Anglia) and the annual rainfall in inches (Aberystwyth). The small circles in the lower part of the figure refer to sites where no grasshoppers were found despite the fact that both soil type and vegetation appeared adequate to support this species. The frequency of B's in the population is indicated by the proportion of the circle shown solid. (Upper part from HEWITT and BROWN, 1970; lower part after Hewitt and John.)

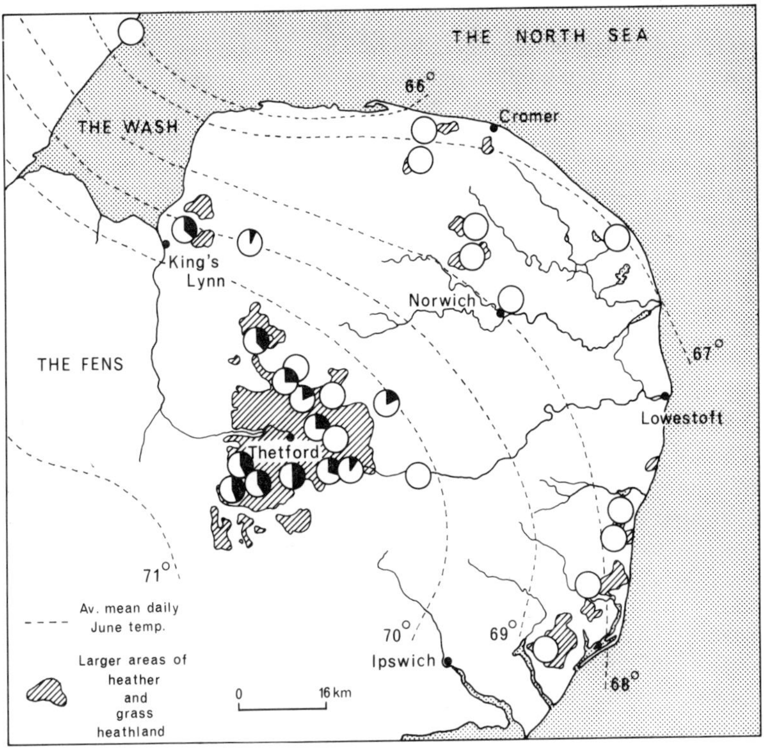
THE NORTH SEA
THE WASH
66°
Cromer
King's Lynn
Norwich
67°
THE FENS
Lowestoft
Thetford
71°
Av. mean daily June temp.
70°
69°
Ipswich
Larger areas of heather and grass heathland
0 16 km
68°

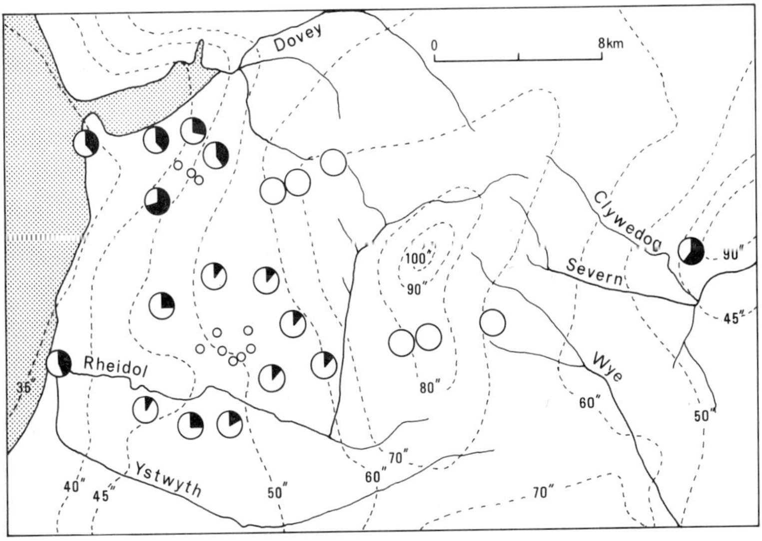
Dovey
0 8km
Clywedog
100"
90"
Severn
90"
45"
Rheidol
80"
Wye
35"
60"
50"
70"
60"
Ystwyth
40"
45"
50"
70"

magnificum, Crystallinum forgetti and *Caltha palustris*. In *Centaurea*, for example, B's are present in all 222 natural populations studied in Scandinavia and Finland though the frequency of plants with B's differs markedly in different areas. In particular there is a marked E–W difference in B frequency, B's are high in the E. and low in the W. Indeed 74 of the 82 populations examined from S.E. Sweden and the Southern interior of Finland have B frequencies averaging 50%. If, therefore, B's are able to maintain themselves at such high frequencies in the absence of any mechanism for numerical increase there must be selection for plants with B's in these areas.

3.2 Structural variation

3.2.1 Inversion systems

(a) PARACENTRIC INVERSIONS The genus *Drosophila* provides the classical case of population variation in respect of paracentric inversion polymorphism and natural populations of many species within this genus consist of mixtures of basic homozygotes, inversion heterozygotes and inversion homozygotes. Rarely, however, do the heterozygotes exceed 50% of the population. Different species reveal different patterns of variation. In some (*pseudoobscura, persimilis, athabasca, nebulosa* and *melanica*) the variability is concentrated in a single chromosome. In others (*willisoni, robusta, azteca, algonquin, sturtevanti* and *paramelanica*) the polymorphism is distributed equally in all the chromosomes. Again many temperate species (*persimilis, athabasca, azteca, robusta* and *pseudoobscura*) have inversions of moderate length but the total number of known inversions is relatively small. On the other hand *D. willistoni* is characterized by a large number of short inversions. *D. prosaltans* also has short inversions but unlike *willistoni*, where inversions are distributed on all the chromosomes, in *prosaltans* they are concentrated on a single chromosome. Inversions are similarly concentrated in *pseudoobscura* where 22 different inversions are known of which 15 are more or less common in natural populations. Here the extent of individual inversions range from 18–60% of the total length of the chromosome concerned (autosome III). Why these differences exist between different species is not known.

With the exception of *Drosophila subobscura* (p. 73) natural populations of *willistoni* contain a greater variety of inversion polymorphism than any other species of *Drosophila*. Some 50 different inversions have been described which are distributed in the five major chromosome arms in the following manner:

XL	XR	IIL	IIR	III	*Total*
11	7	8	6	18	50

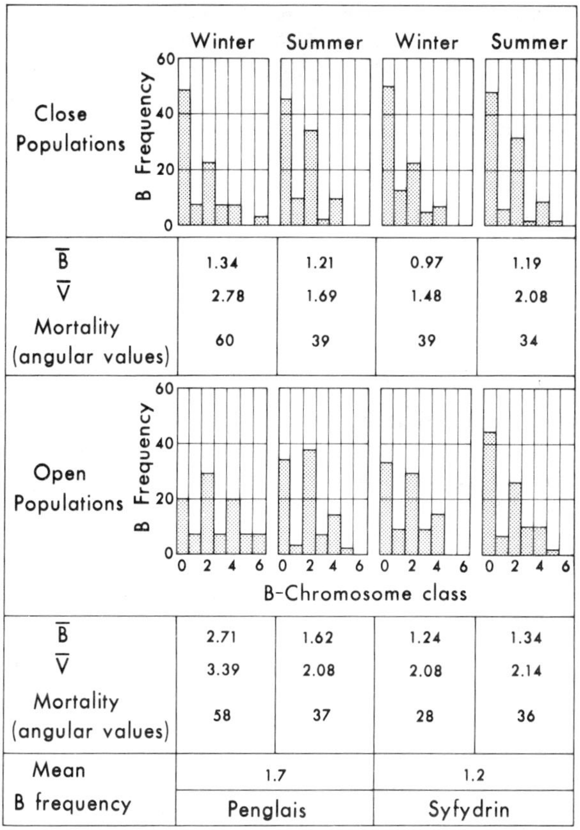

Fig. 3–6 Differential survival of rye plants with B-chromosomes following variable selection in close and open experimental populations sown at two sites (Penglais and Syfydrin) in Aberystwyth. (After Rees and Ayonoadu.)

Here chromosome III, the longest member of the complement, has the most inversions but the second largest number is found in the shortest arm (XL) and the smallest number of inversions is found in the second longest arm (IIR).

Dobzhansky has distinguished between flexible and rigid polymorphisms. In the first category are those cases where the frequency equilibria of the polymorphism vary with latitude, altitude, season or temperature and we shall meet examples of these in Chapter 4. Polymorphisms of this type form a characteristic feature of many of the species of *Drosophila* which are endemic to particular geographical areas including *pseudoobscura, paramelanica, persimilis, flavopilosa* and *robusta*. By contrast, geographically widespread species (*repleta, simulans* and *virilis*) tend either to be monomorphic throughout or else are characterized by rigidly maintained polymorphisms which do not follow ecogeographical gradients. Indeed of the cosmopolitan species, only one, *funebris*, shows a pattern of polymorphism resembling that of the endemic species. While single inversions lead to an organization of gene complexes protected against dissociation (p.17), integration of the genotype can go even further than this. Some independent inversions occur together even when the possibility exists of recombination between them. In these cases crossing-over between such linked inversions is much lower than would be expected on the basis of the distance between them (see p. 71).

Dobzhansky has also argued that inversions in natural populations are adaptively important because they protect co-adapted complexes of polygenes. This concept of co-adaptation implies selection both for particular alleles at different loci (epistasis) and for different alleles of the same locus between different inversions (heterosis). Presumably such co-adaptation results from the interaction of polygene complexes but the details are not known. One of the principal lines of evidence used to support co-adaptation comes from the observation that alternative gene arrangements reach characteristic stable equilibrium frequencies in laboratory populations when the arrangements are derived from the same natural population but not when they come from different populations.

(b) PERICENTRIC INVERSIONS By contrast with paracentric inversion systems far less is known about pericentric inversions though polymorphisms with respect to them have been recorded in a number of rodent and grasshopper species. All black rats, *Rattus rattus*, collected in E. and S.E. Asia are characterized by a diploid number of 42 and all the subspecies found in this area show essentially similar karyotypes with 13 pairs of telocentrics (autosomes 1–13), 7 pairs of small metacentrics (autosomes 14–20) and an acrocentric sex chromosome pair. Pairs 1, 9 and 13, however, show a telo-acrocentric polymorphism whose extent varies in different localities. The situation in chromosome 1 has been studied in some detail in Japan. Here T/T karyomorphs are found in the N. and N.W. areas whereas the T/A and A/A types are found together with T/T types in S. and S.E. Japan (Fig. 3–7). Moreover in the latter group the frequency of A-chromosomes varies in different populations and is highest in Eastern populations. The reasons for these differences are not known though the climate in N. and N.W. Japan is certainly more severe than in the S. and E.

Pericentric inversions also form an important part of the genetic system in Morabine grasshoppers. For example in *Keyacris scurra* inversions occur in the CD and EF chromosomes (Fig. 3–8) and these are known to interact with one another in maintaining a stable equilibrium. White has argued

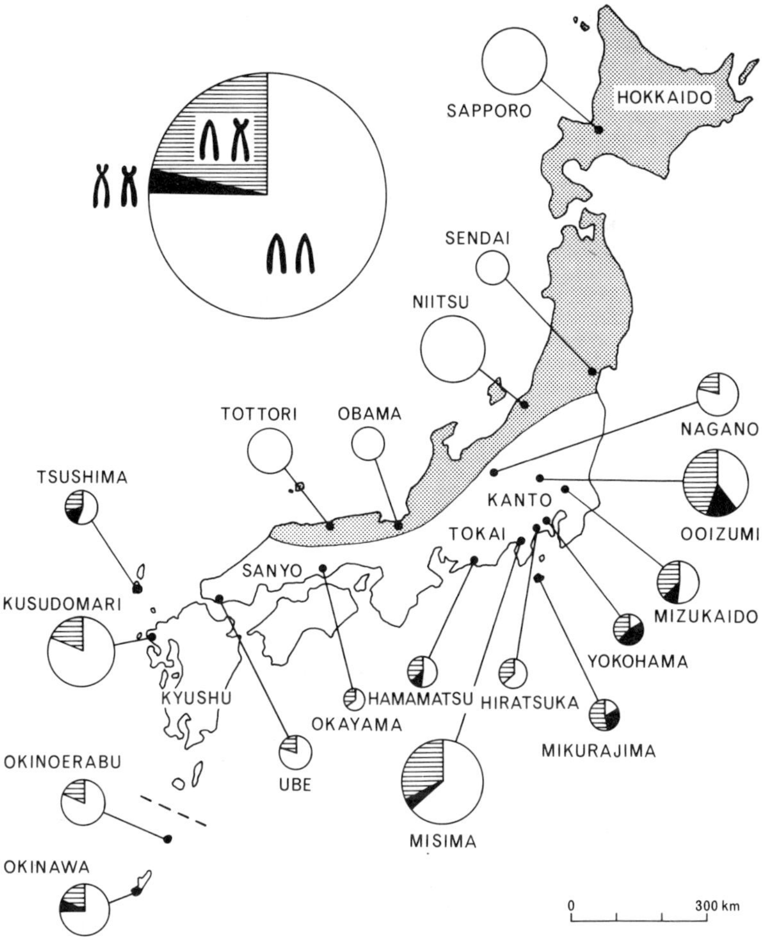

Fig. 3–7 Frequency distribution of polymorphism for a pericentric inversion in chromosome-1 in Japanese populations of the black rat, *Rattus rattus*. The relative size of the population sampled is indicated by the size of the circle used to denote that population. (From YOSIDA, TSUCHIYA and MORIWAKA, 1971.)

that the standard sequences of CD and EF are size increasing while the *Bl* and *Td* inversion sequences are size decreasing so that there appears to be a different combination of '+' and '–' polygenes for size determination in the different cytological sequences found in this species. White also assumes that the large and small phenotypes in the existing populations occupy different ecological niches at least for some part of their life cycle and that the selective values of the various cytotypes are frequency dependent. As yet, however, there is no evidence to support either assumption.

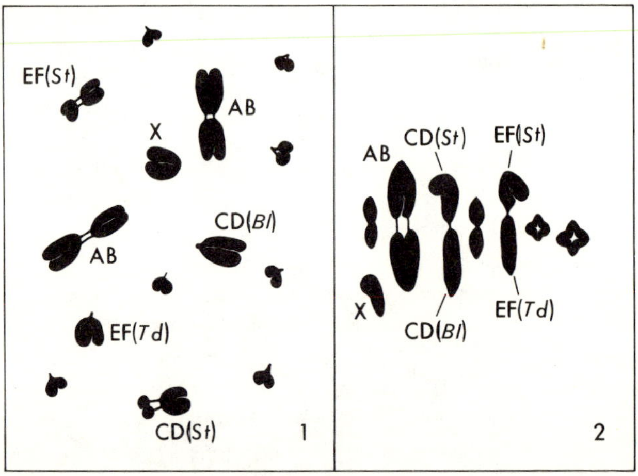

Fig. 3–8 Mitotic (1) and meiotic (2) metaphase plates of a male individual of the 15-chromosome race of *Keyacris scurra* heterozygous for pericentric inversions in the CD and EF chromosomes. In each case *St* represents the standard condition while *Bl* and *Td* represent the Blundell and Tidbinbilla inversions respectively. (2, after White.)

3.2.2 Interchange systems

Interchanges are without question the least common of all structural rearrangements found in natural populations and successful interchange systems are known in only a few plants and in still fewer animals. Interchanges differ in another respect from other structural types. While they may float in populations in a polymorphic state they may also become fixed in a population not only in the structurally homozygous state (translocation races) but also in the heterozygous state (permanent structural heterozygotes).

Among plants they occur especially in the Onagraceae of N. America. Here they are best known in the subgenus *Oenothera* where the precise population structure varies with geographical location. Some 438 complexes have been analysed and these fall into 10 groupings. Each of these tends to have a definite range, distinctive phenotypic consequences and, coupled with this, distinctive cytogenetic characteristics. Plants in Mexico and Central America and those resident in California and contiguous areas have large and open-pollinated flowers and form seven bivalents at meiosis. In Arizona, New Mexico, Nevada and Utah the members of this subgenus, although still large flowered and open-pollinated, commonly show small ring multiples; these plants do not breed true for the interchanges however. From the Rocky Mountains eastward, populations are made up almost entirely of a plethora of small flowered, self-pollinated and true breeding lines, each with a single ring multiple of fourteen chromosomes at meiosis and a system of balanced lethals which allows these plants to breed true. Because the interchange rings and chains in *Oenothera* orient on the first meiotic spindle in such a way that alternate chromosomes in the multiple move to opposite poles, only two types of gamete are consistently produced in any race (Fig. 3–9). The seven chromosomes which enter any one gamete therefore carry one of two particular parental arrangements which are termed Renner complexes. Although each complex in fact contains seven chromosomes they behave essentially as though they constitute a single linkage group. Additionally the differential segments of each complex carry a system of balanced lethal factors which result either in the death of zygotes containing two identical complexes (zygotic lethals) or else the death of

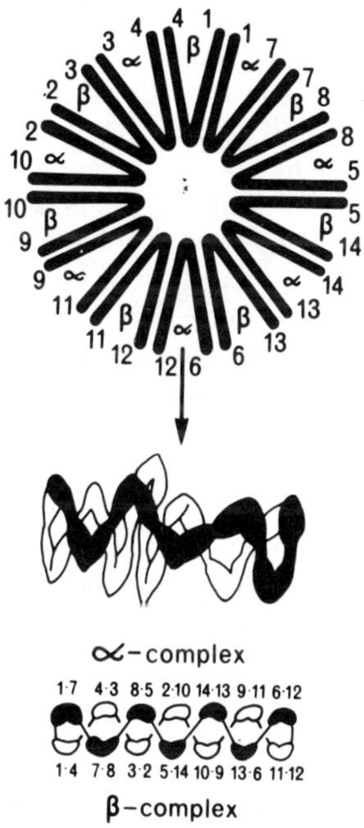

Fig. 3–9 Meiotic chromosome behaviour of a race of the evening primrose (Oenothera) from Colorado, USA. The 14 chromosomes which are present in the diploid set of this plant form a giant ring of fourteen members in which alternate members come from one or other of two distinct complexes (α and β Skyview) which regularly segregate from one another at first anaphase. (After Darlington.)

one of the two particular complexes at the megaspore stage and the death of the corresponding complex at the microspore stage (gametic lethals). Megaspore lethals are also associated with a system of megaspore competition, known as the Renner effect, which guarantees that in every ovule the single megaspore which functions as the egg nucleus carries a pollen lethal.

In the case of zygotic lethals only one type of zygote survives, while with gametic lethals, which are the most common, 50% of the eggs and pollen produced die but seed set is almost perfect. The lethals either prevent outcrossing or else reduce it to a very low level. As a result of this all individuals within a strain tend to be identical genetically both with their parents and with each other. There is, therefore, little or no variation within a strain and the variation in nature is due to the many thousands of strains or lines that exist. Two of the eastern groupings only are exceptional. Both are small and consist of bivalent-forming plants with large, open-pollinated flowers and free of lethals. One of these, the grandifloras, exists in the vicinity of Mobile, Alabama. The other, the argicollas, is distributed along the Appalachians from Pennsylvania to Virginia, primarily on shale barrens.

Apart from *Oenothera,* interchanges are known also in natural populations of five other genera of Onagraceae—*Clarkia, Stenosiphon, Cammisonia, Gayophytum* and *Gaura.* The situation is particularly impressive in *Clarkia* where translocation heterozygotes have been found in 14 of the 34 known diploid species which are endemic to the N.W. of N. America and in the single polyploid complex found in the S.W. of S. America. Heterozygotes are particularly common in *unguiculata, williamsonii, amoena* and *dudleyana* though the frequency varies greatly among populations, ranging from 0–69% in *unguiculata* for example. In all these cases the interchanges have involved the major portions of chromosome arms and disjunction is almost always alternate though the ring multiples formed are mostly small. By contrast the genus *Gayophytum* comprises a small homogeneous group of species endemic to Western N. and S. America. Six of the species are diploid ($x=7$) and three are tetraploid. One of the six diploids, *G. heterozygum* regularly forms a ring of fourteen at meiosis whereas the remaining diploids form pairs only or at the most rings of four or six. In *G. heterozygum* ring formation is associated with abortion of about half the ovules and pollen grains. The species is thus a complex heterozygote maintained by balanced lethals.

Complex hybridity has also evolved in *Isotoma petraea* (Lobeliaceae) an Australian species which is found in rocky exposed situations. Because of this habit-preference the species occurs as a series of small, well-isolated populations. In the greater part of its range, populations of *petraea* carry a proportion of interchange heterozygotes ranging from 10–50% of the population. At the extreme S.W. of its range, however, local populations are composed of plants in which cross pollination has largely been replaced by autogamy and which are all multiple interchange heterozygotes with a marked tendency for the ring size to increase along a N.E.–S.W. cline. The complex hybrids breed true by selfing since they are stabilized by a balanced lethal system operative at the zygotic level. In plants heterozygous for a single interchange, disjunctional anaphase-I configurations occur in about 90% of the cells but the frequency of non-disjunction increases with increasing ring size (Table 5). This striking loss of meiotic regularity with increasing ring size is paralleled by a corresponding decrease in pollen fertility and post-meiotic ovule abortion. Seed production is thus drastically reduced in these complex heterozygotes.

Table 5 Meiotic behaviour of interchange heterozygoes in *Isotoma petraea* (data of JAMES, 1970) ⊙ = a ring of.

Population	*Various* ⊙ 4	*Pigeon Rock* ⊙ 6	*Berring-booding Rock* 2 ⊙ 6	*Muntagin* 2 ⊙ 6	*3 mile Rock* ⊙ 10	*Bencubbin* ⊙ 12	*Merredin* ⊙ 14	*Mt Stirling* ⊙ 14
% Disjunction at AI	90.2	58.8	65.0	46.8	29.5	23.2	22.5	16.3
% Pollen fertility	—	73.6	—	37.2	57.9	48	34.6	—

It is clear from these cases that, in nature, interchange heterozygosity is commonly established and maintained in conditions where inbreeding is imposed on colonizing derivatives of normally outbred species. Under such circumstances the role of interchange is to facilitate the conservation of the allelic heterozygosity of the outbred form in the face of inbreeding.

Interchange heterozygosity is even less common in animals than in plants and only two cases are on record—one in cockroaches, the other in scorpions. The situation is best known in the American cockroach, *Periplaneta americana*, where all the wild populations examined to date have been found to contain 50% or more individuals heterozygous for rings of IV, VI or rarely VIII (Table 6). Although the rings are mainly

Table 6 Incidence of interchange heterozygotes in wild populations of the American cockroach, *Periplaneta americana*, from Britain and Pakistan (pooled data of JOHN and LEWIS, 1968 for Britain, JOHN and QURAISHI, 1964 for Pakistan).

Populations	*Homozygotes*	*Heterozygotes* 1 IV	2 IV	3 IV	1 VI	1 VI + 1 IV	1 VIII	*Total individuals*	*% hetero-zygotes*
Britain	33	56	25	3	9	5	—	121	72
Pakistan	21	8	5	—	6	5	1	46	50
Totals	54	64	30	3	15	10	1	167	66.4

small, at least 8 different interchanges must have occurred in the 16 autosomes that make up the diploid complement. In these wild populations, as opposed to those maintained in culture, the degree of maximum multiple formation is high for all karyotypes (> 70% in a majority of cases). Likewise the frequency of disjunctional orientation is high for all multiples (> 80%).

3.2.3 Whole arm transpositions

It was Robertson who first suggested that two rod-shaped non-homologous chromosomes might fuse together to form one V-shaped metacentric. He also suggested that the reverse process might occur so that two rods could be derived from one metacentric. It is not clear from Robertson's publications how he envisaged these processes to occur. We now know that 'fusion' can occur in at least one of two ways (Fig. 2–4). Thus if a break occurs in the short arm of one acrocentric chromosome and in the long arm adjacent to the centromere of the other a metacentric would result (whole arm transposition). Coincident with this, union might also occur between the acentric short arm section and the small centric piece to give a small metacentric, thus $A + A \rightarrow M + m$. Although such an m-chromosome can persist it is usually lost in natural populations presumably because of its inability to regularly form chiasmata. In some cases, of course, it may float in natural populations as a supernumerary chromosome. Alternatively if the short arms of both acrocentrics were broken at the margin of the centromere with reunion at the edge of each centromere this would also yield a metacentric. So would breakage and reunion within the limits of each centromere (true centric fusion). Past interpretation of such Robertsonian systems has been prejudiced by the assumption that all of them are indicative of whole arm transposition and many of the Robertsonian relationships in animals have been interpreted this way. There is, of course, the additional complication that, unless one has a convincing basis for distinguishing the direction of chromosome evolution, it is usually difficult to decide between fusion or fission. There are therefore few well defined instances of whole arm transposition on record. One of the most convincing is that recently reported in *Gibasis schiedeana* (Commelinaceae).

Studies of Mexican collections of *G. schiedeana* have demonstrated the existence of two cytotypes. These are indistinguisable morphologically in their aerial parts but are readily distinguished in cytogenetic structure. One is a self-sterile diploid with $2n = 2x = 10 = 4M + 6A$; the other is a self-fertile autotetraploid with $2n = 4x = 16 = 12M + 4A$ in which the A chromosomes are visibly and clearly genuine acrocentrics. The difference in breeding system and in ploidy level both indicate that the direction of change must have been $x = 5$ to $x = 4$ and a comparison of the haploid genomes suggests that the tetraploid ($x = 3M + 1A$) has been derived from that of the diploid ($x = 2M + 3A$) by a single centric fusion. This has been confirmed by a study of F_1 hybrids between the diploid and the tetraploid. These triploids have $2n=3x=8M+5A$. Six of these metacentrics and three of the acrocentrics have not been involved in the transposition. They can be distinguished by their pairing behaviour at meiosis and frequently form trivalents. The remaining two metacentrics and two acrocentrics give a maximum multiple of four with the structure acro-meta-acro-meta thus confirming the occurrence of a genuine whole arm transposition (Fig. 3–10).

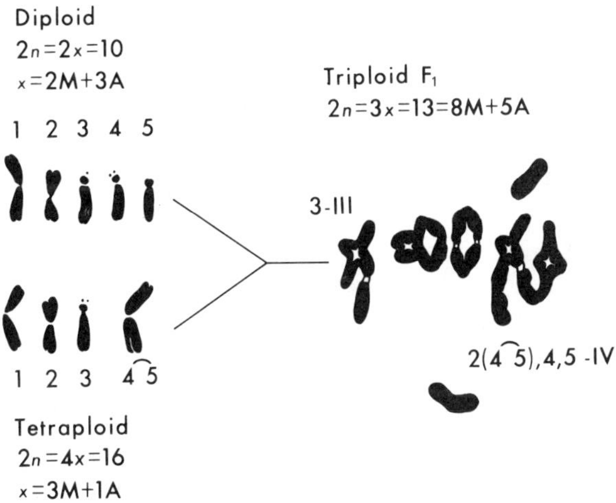

Fig. 3–10 The relationship of the diploid and autotetraploid cytotypes of *Gibasis schiedeana*, a plant species native to Central America. (After Jones.)

3·3 Structural/Numerical variation

3.3.1 Fusion/Fission systems

(a) FUSION The black rat, *Rattus rattus*, occupies a wide range of habitats throughout the world and can be subdivided into many different subspecies in terms of its morphological characteristics. All black rats in East (Japan, Korea, Formosa and Hong Kong) and S.E. Asia (Philippines, Thailand, Malaysia and Indonesia) are characterized by a diploid number of 42. By contrast rats collected in Oceania (Australia, New Zealand and New Guinea) have $2n=38$ (Fig. 3–11). This difference depends on the presence in Oceanian forms of two large metacentric pairs. One of these is equivalent to a 4⌒7 fusion while the other represents an 11⌒12 fusion (Fig. 3–12). This mode of origin of fusion metacentrics has been confirmed by an analysis of G-banding patterns. Equivalent rats have also been found in Argentina, Texas, Hawaii, Rome and Cairo. Thus it seems likely that, during the migration of black rats from the presumed centre of

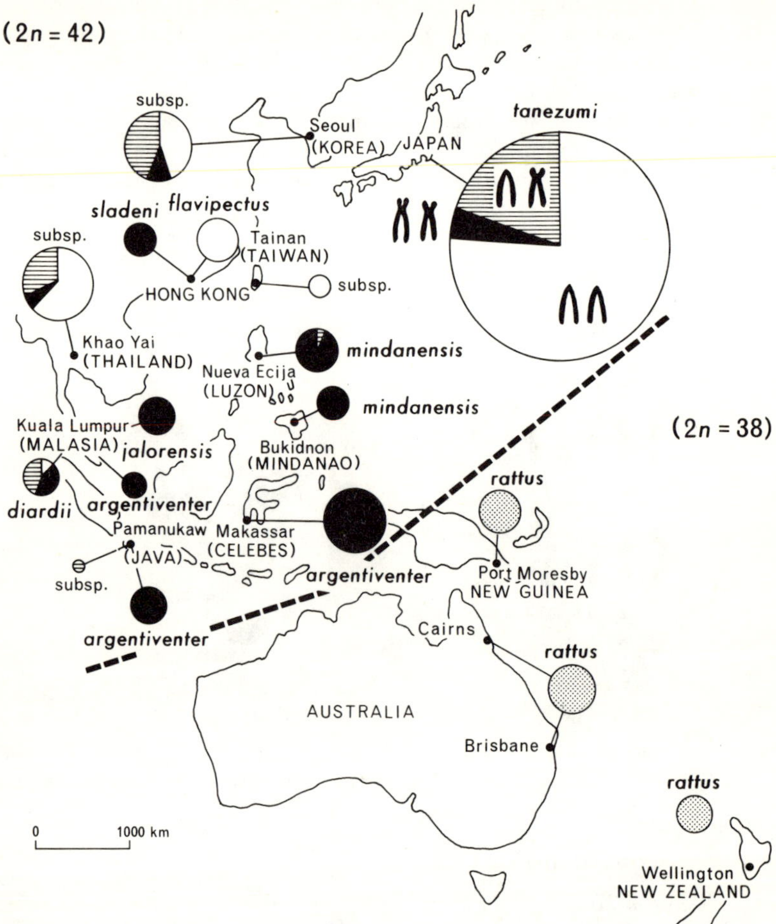

Fig. 3–11 Chromosome variation in the subspecies of *Rattus rattus* found in eastern and south-eastern Asia and in Oceania. In cases where the subspecies has yet to be named the population is shown simply as 'subsp.'. The relative size of each population sampled is indicated by the size of the circle used to denote that population. (From YOSIDA, TSUCHIYA and MORIWAKA, 1971.)

origin of the species in the Indo-Malayan region through S.W. and central Asia to Europe, pairs 4 and 7 and 11 and 12 respectively underwent fusion so that $2n = 38$ type rats were distributed first to Europe, from where they have spread secondarily to N. and S. America on the one hand, and to Oceania and Africa on the other, with the commercial movements of man. In keeping with this view the Asian type ($2n = 42$) is distributed in northern parts of India and in Pakistan and Nepal. The Oceanian type is found in the southern parts of India, in Pakistan and also in Central Asia. Hybrids between the two types ($2n=39$) occur in Karachi while rats from Ceylon show $2n=40$ with a single $\widehat{11\ \ 12}$ fusion.

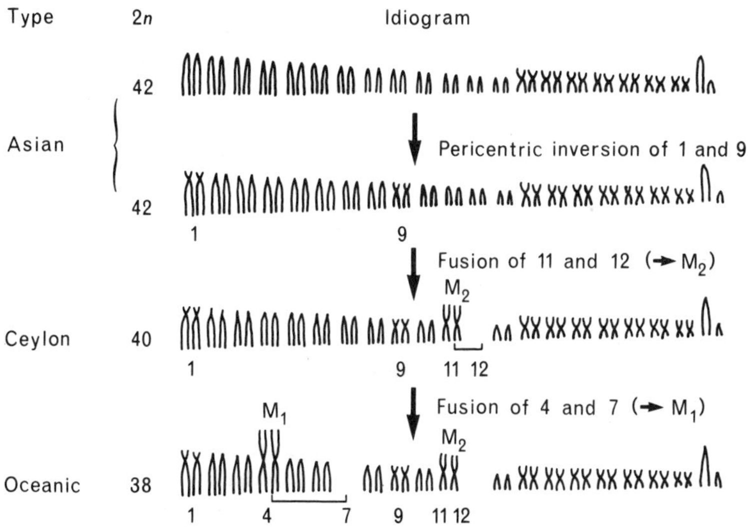

Fig. 3–12 The presumed relationship of karyotypes in Asian and Oceanian types of *Rattus rattus*.

(b) FISSION Two karyotypically distinct races of the common shrew, *Sorex araneus*, have been identified in Europe.

(i) Race A occupies lowland W. Europe and has $2n=23$ (XY_1Y_2♂) in which all the members of the karyotype are two-armed entities.
(ii) Race B has an alpine, N. and E. distribution and shows fission polymorphism in from 1–6 chromosomes which are present either as metacentrics or as two equivalent telocentrics (Fig. 3–13). This leads to a variation in the diploid number from 20–32. Polymorphic populations have been identified from the Swiss and French Alps, in Britain, Scandinavia, W. Germany and E. Poland.

Race B differs from race A by three pericentric inversions and one tandem translocation (Fig. 3–13). Zones of overlap of the two races have been found in Switzerland without any evidence of hybridization so that the two races can be regarded as cryptic species.

3.3.2 Supernumerary segment systems

Populations of grasshoppers frequently show differential poly-

UNIVERSITY OF BRISTOL VET. SCIENCE

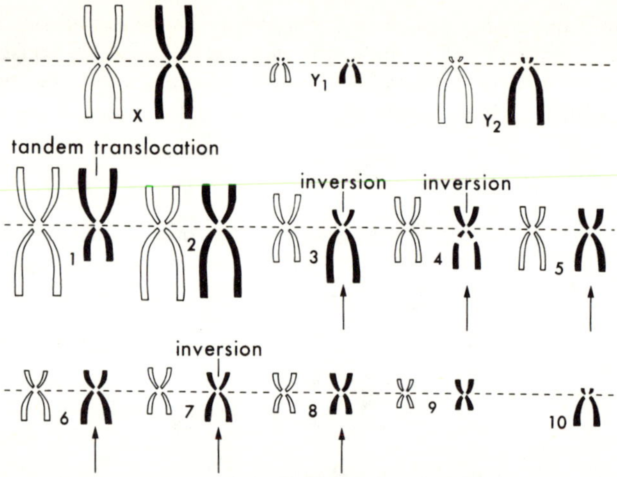

Fig. 3–13 Chromosome variation in the common shrew, *Sorex araneus*. Two races (A and B) of this animal are known which in all probability represent distinct species. The chromosome on the left of each pair (shown in outline) represents the condition in race A. The two races differ by one tandem translocation (=whole arm translocation) involving chromosome 1 and 10 and three pericentric inversions involving chromosomes 3, 4 and 7. Additionally in race B six of the chromosomes (3, 4, 5, 6, 7 and 8, arrowed) may be involved in Robertsonian polymorphism. (From FORD and HAMERTON, 1970, by permission of the Zoological Society of London.)

morphism with respect to the presence and frequency of small supernumerary segments on one or more members of the standard set (Fig. 3–3). These are always at least partly heterochromatic in character and while they may occupy an interstitial region of the chromosome they are more usually proximal or distal in location, which coincides with the most common siting of heterochromatic blocks in the species concerned. They are particularly well known, and especially extensive, in populations of *Chorthippus parallelus* (the meadow grasshopper). Here all populations so far examined in Britain and Europe are regularly polymorphic for segments on the two smallest members of the complement. Additionally in Unterloibel, in Austrian Karynthia, a segment is also present on the M_6 chromosome. The segment system here is therefore widespread and presumably of considerable antiquity.

Equivalent segments occur also in plants. The most extensive of these so far known is found in *Scilla sibirica* where the heterochromatic regions are of two kinds:

(a) pro-centric and showing enhanced quinacrine fluorescence, and
(b) interstitial and terminal and showing reduced quinacrine fluorescence.

After denaturation and re-annealing treatment both types stain with Giemsa and with acetic orcein, the latter tending to stain preferentially those segments showing reduced fluorescence. An analysis of chromosome variation in 20 plants taken from a single population has shown that all of them are unique in their heterochromatic segment structure and all six chromosomes of the haploid complement are polymorphic (Fig. 3–14). The centromeric bands are constant but all other bands are variable. Moreover extra heterochromatic regions may arise in addition to, or else in the place of, euchromatic material. This variation gives rise to differences in the overall heterochromatin content ranging from 11–20% of the haploid complement.

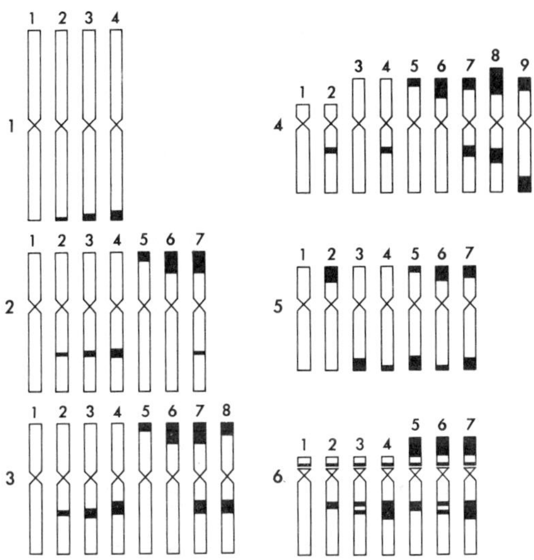

Fig. 3–14 G-band variation in a single population of the bluebell *Scilla sibirica*. Six chromosome types are present in this species and each occurs in a number of variant forms. Chromosome 1 is the least polymorphic (4 variants) and chromosome 4 the most (9 variants). Chromosomes 2, 5 and 6 have 7 variants and chromosome 3 has 8. (From VOSA, 1973.)

4 Chromosome Variation in Natural Populations: II. Spatio-Temporal Patterns of Variation

To everything there is a season
And a time to every purpose

Ecclesiastes 3:1

4.1 Eco-geographical variation

Two major patterns of eco-geographical variation are commonly encountered in natural populations.

4.1.1 Clinal variation

Some features of the environment change in an abrupt fashion so leading to sharp ecological boundaries. Others change gradually with distance. The cytogenetic structure of a series of populations belonging to a given species can sometimes be shown to exhibit a related unidirectional change across a geographical transect.

One of the clearest demonstrations that inversion polymorphism in *Drosophila* is adaptive comes from the fact that populations may exhibit a graded series, or cline, in the distribution of particular inversions which corresponds to an obvious environmental gradient. *Drosophila flavopilosa*, for example, is a neotropical species whch is widely distributed in S. America. It is also ecologically restricted since in its pre-adult stage it exploits a precise niche made available by a single species of Solanaceous shrub, *Cestrum parqui*, which ranges from sea level to about 1500 m in the pre-Andean region. The only variation found in populations from central Chile involves paracentric inversions in the basal two-thirds of the right arm of chromosome *V*. Here 4 distinct inversions are known and they show marked variations in frequency in different populations. The two most abundant inversions show an altitudinal gradient in their distribution. Thus in two valleys near Santiago, Chile, which run from the Andes to the Pacific Coast, heterozygotes for the A inversion are more frequent at high altitudes than at sea level while, by contrast, heterozygotes for the *B*–inversion are more abundant at sea level (Fig. 4–1). Likewise in *D. robusta* inversions X*L*–1, 2L–3 and 3R decline progressively in frequency as one goes South. Indeed the clines in these arrangements are perhaps the most striking in the species and are gradual over long distances.

In the subspecies *Thomomys bottae grahamensis*, a pocket gopher which inhabits desert mountain ranges, populations from the Graham (=Pinaleno) mountains of S.E. Arizona are chromosomally polymor-

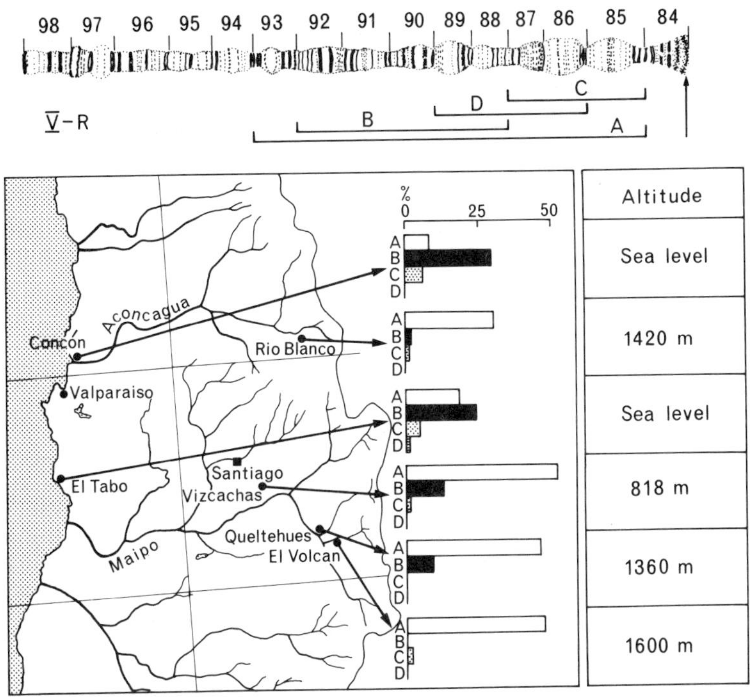

Fig. 4–1 Altitudinal variation in the percentage frequencies of 4 inversions found in a heterozygous state in chromosome V–R of *Drosophila flavopilosa* from two valleys near Santiago, Chile. (From BRNCIC, 1962.)

phic. While the diploid number remains constant at 76, different numbers of telo- and metacentric elements occur within different individuals. It is not yet known whether this results from pericentric inversion; an alternative possibility is that supernumerary heterochromatic arms are present on the telocentrics in some populations so converting them into metacentrics as in *Peromyscus maniculatus* (see below). Whatever the precise situation turns out to be there is a clear altitudinal variation in the karyomorph composition of particular populations. Beginning at about 2150 m the modal number of telocentrics increases with elevation (Fig. 4–2). Other desert ranges in the Chiracahua, Huachuca and Santa Catalina mountains do not show an equivalent intra-population karyopolymorphism. Here there is certainly a clear distinction between populations at the base of these mountains and those at the top but this is abrupt and not clinal. Significantly these other ranges do not reach such a high elevation and they show considerably less diversity of habitats and micrographical differentiation.

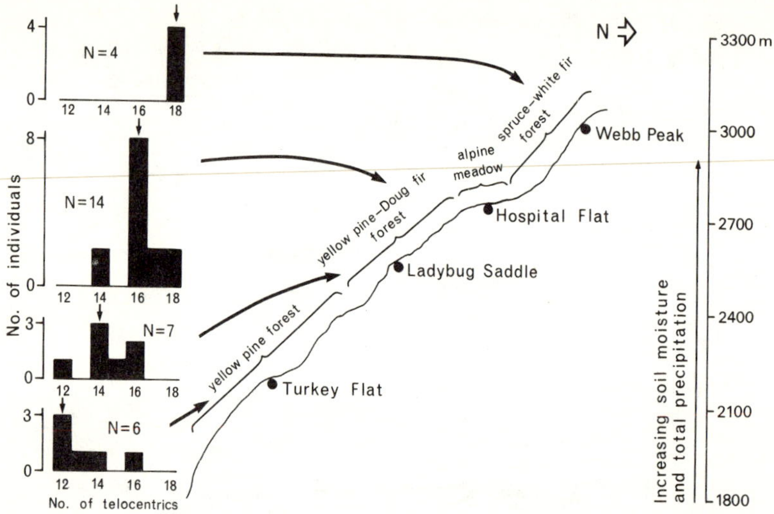

Fig. 4–2 Clinal variation in the number of telocentric chromosomes present in four populations of the pocket gopher *Thomomys bottae grahamensis* taken from the south-facing slope of the Graham mountains, south-eastern Arizona. (From PATTON, 1970.)

Two major ecological forms have developed in *Peromyscus maniculatus*, a grassland form and a forest form, and within each of these forms numerous subspecies exist. The number of biarmed chromosomes among the subspecies ranges from 16 to 42 and involves the addition of heterochromatic segments to the C-bands present at the centromere of all rod-members of the complement. Clinal tendencies exist for the number of biarmed chromosomes to increase in cooler climates as a function of either elevation (*P.m. orcas*) or latitude (*P.m. gambelli*).

Perhaps the most impressive example of a relationship between population structure and microgeographical conditions is that found in the littoral species *Nucella lapillus* (the dog-whelk). In the vicinity of Roscoff (Brittany, France) this marine gastropod exhibits a variety of chromosome forms with diploid members varying between 26 and 36. Eight pairs of metacentrics are always present. The 26-chromosome form has an additional five metacentric pairs which are replaced by ten acrocentric pairs in the 36-chromosome form. In addition to these two homozygous states, heterozygotes combining all possible intermediate conditions are also found with from one to five 1M/2A trivalents at meiosis. Non-disjunction occurs only infrequently in them so that all the intermediates are viable and interfertile.

Different populations show different frequencies of the possible karyomorphs. Moreover the cytogenetic structure of particular populations is correlated with the ecological character of the habitat. Monomorphic $2n=36$ populations are confined to sites sheltered from heavy wave action. Monomorphic $2n=26$ populations occur principally in exposed sites though they are sometimes found on sheltered shores too. Finally, in localities where the $2n=36$ form occurs at all, populations with intermediate chromosome number occupy shores intermediate in exposure.

In southern Britain a majority of populations between Milford Haven and the Straits of Dover prove to be monomorphic $2n=26$ despite the wide differences in the degree of wave action and exposure which obtain over this area.The only polymorphic populations obtained are those from the major bays of the south-west. Acrocentrics reach particularly high frequencies on the Dorset coast and in South Devon. In both of these areas there appears to be a negative association between the frequency of acrocentrics and aspects of water movement including not simply wave action but also tidal currents. Thus high chromosome number appears to be favoured where total water movement is least.

4.1.2 Central–peripheral variation

In many of the species of *Drosophila* where paracentric inversion polymorphism has been studied there is a decline of the polymorphism in peripheral populations. This is particularly true in *robusta* and *willistoni*, where the amount of inversion heterozygosity falls dramatically at the periphery of the species range. Two quite different explanations of this pattern of variation have been offered. On the one hand it has been argued that the polymorphism present in the central populations reflects the exploitation by such populations of a wide variety of ecological niches (niche-width hypothesis). Conversely since peripheral situations offer few niches this is reflected in the relative cytogenetic uniformity of population structure. One of the main difficulties with this hypothesis is that it is difficult to quantify niche-width in a meaningful manner because so little is known about the ecology of *Drosophila*. Indirect support for this concept comes from comparisons between closely related species which differ in their pattern of distribution. Thus the common and widespread species *D. polymorpha* has more heterozygous inversions per individual than its closest relative *D. cardinoides*. Similarly the common species *D. guaramunu* has more inversions than its less frequent relative *D. griseolureata*. Again in *D. nebulosa*, those populations which inhabit the exacting environment of the desert show fewer inversions than do the inhabitants of the richer and more diversified rain forest and savanna environments.

Even so several reservations have to be made in applying this seemingly simple hypothesis. Thus the luxuriant forests on the coasts of the states of Bahia and Espiroto Santo in Brazil, where *D. willistoni* is the dominant species of its group, are inhabited by populations which show almost as little polymorphism as those of the adjacent desert regions. Again *D. montana*, which is ecologically very restricted, has more inversions than *pseudoobscura*. Finally cage populations of *Drosophila*, which experience a much more uniform environment than any found in nature, usually conserve the inversion polymorphism characteristic of natural populations.

The alternative explanation argues that, since inversion heterozygotes restrict recombination, peripheral homoselected populations will be expected to have higher rates of recombination than will central heteroselected populations. In *D. robusta* Carson has expressed the degree of heterozygosity in terms of an index of free crossing-over. This index represents the percentage of observable euchromatin in polytene chromosomes which is free of inversions and in which, therefore, crossing-over is unhampered. In these terms it can be shown that this index is low in populations from the centre of the geographical range of the species but increases as more peripheral conditions are approached (Fig. 4–3). Notice that this in no way proves that a real difference exists in the recombination rates of the populations involved since there is no precise information on what effects, if any, the inversions have on the

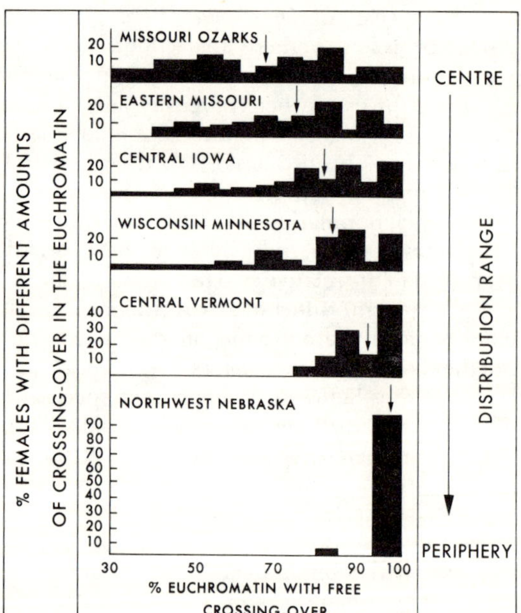

Fig. 4–3 Differential patterns of crossing-over in natural populations of *Drosophila robusta*. The arrows indicate the mean of the distribution in each case. (From CARSON, 1958.)

amount of recombination in the linked, non-inverted regions or whether inter-chromosomal effects are operative in this case (p. 72). Indirect evidence in support of this hypothesis is, however, provided by the fact that peripheral populations also respond to selection both faster and to a greater extent than those from the central area.

What appears to be a comparable case of a central–peripheral difference is found also in *Trimerotropis helferi*, a trimerotropine grasshopper. Trimerotropines are grasshoppers of arid regions and this genus, which is endemic to N. America, includes a number of species which are polymorphic for pericentric inversions. Such a polymorphism exists in *T. helferi* a species of restricted distribution which is confined to the coastal region of N. California and S. Oregon. Here the eighth autosome in order of size (M_8) exists either as a telocentric or as a submetacentric. Three types of individual are thus found—8^t8^t, 8^t8^m and 8^m8^m. In 8^t8^m heterozygotes pairing is non-homologous, the two chromosomes pairing straight with no reverse looping. Because of this there can be no crossing-over within the limits of the inversion and the species avoids the fertility handicap inherent in the adoption of pericentric inversion. A comparative analysis of six populations of this limited species covering its entire range of distribution indicates that inversions are absent from the northernmost periphery (Table 7).

Table 7 Population characteristics of the grasshopper *Trimerotropis helferi*, a coastal species of N. California (data of SCHROETER, 1968).

Population		*% 8^t8^m heterozygotes*	*Mean Chiasmata/Cell* 8^t8^t	8^t8^m	8^m8^m	*Mean cell Chiasma frequency per population*
N. periphery	Bandon	0	—	—	—	—
	Orick	45	22.18	24.11	24.07	23.45
	Orcata	38	21.95	23.85	24.33	23.37
	Petulia	49	—	—	—	—
	Cleone	52	20.68	22.57	23.15	22.13
S. periphery	Point Arena	32	—	20.98	21.28	21.13

4.2 Seasonal variation

In Israel the mantid *Ameles heldreichi*, which has two generations per year, is found in three karyomorphs. By comparing spring and autumn generations of a Jerusalem population it has been possible to show that the two samples differ markedly from one another in karyomorph frequencies suggesting different adaptive values of the various cytotypes at different times (Table 8).

Regular temporal fluctuations in the frequency of particular inversions has also been found in specific populations of *D. pseudoobscura, D.*

persimilis, D. robusta, D. funebris and *D. melanica*. The cyclical nature of these changes suggests that the fluctuations are an adaptive response to seasonal changes though it is difficult to identify the particular environmental factors responsible.

The most extensive data on seasonal changes in chromosome frequencies are unquestionably those relating to populations of *D. pseudoobscura* from northern California. For example at Pinõn Flats,

Table 8 Seasonal variation in the frequency of chromosome types in males of *Ameles heldreichi* from Jerusalem (data of WAHRMAN, 1954).

	Male constitution						
	$2n=27$ (2 V + 25 T)		$2n=28$ (1 V + 27 T)		$2n=29$ (29 T)		
Season	No.	%	No.	%	No.	%	*Totals*
Spring 1952	16	47·06	18	52·94	—	—	34
Autumn 1952	22	73·33	6	20.00	2	6.67	30
Early Spring 1953	32	57·14	23	41·07	1	1·79	56
Spring 1953	34	46·58	36	49·32	3	4·11	73
Totals	164	—	83	—	6	—	193

which lies on the desert-exposed slope of Mount San Jacinto, California, at an elevation of 1200 m, the three commonest third-chromosome variants are known as Standard (ST), Arrowhead (AR) and Chiricahua (CH). Despite the unexplained fact that the frequencies of these variants change from year to year there are consistent cyclical fluctuations which have remained essentially the same over the thirty years this population has been studied (Table 9). Specifically, within any one year the frequency of ST tends to be high early in the season but decreases in spring and then recovers again during summer and autumn. The frequencies of AR and CH reverse this pattern. The fact that the same kind of changes are repeated annually suggests strongly that different gene sequences are favoured at different times within a given season. But what aspects of the seasonal environment lead to increases and decreases of the fitness of particular karyomorphs is not known.

Equivalent seasonal changes do not occur in all populations of *D. pseudoobscura* on Mt. San Jacinto though they have been found at Andreas Canyon which lies at the edge of the desert at an elevation of 240 m. Some populations, like that at Keen Camp on the N. side of the mountain at an elevation of 1290 m, show no changes of any sort. Still others show distinct patterns of fluctuation. For example at Mather, located N. of San Jacinto on the western slopes of the Sierra Nevada at an elevation of 1200 m, only ST and AR are involved in seasonal oscillation, the frequency of ST increasing steadily from early to late collections with AR declining

Table 9 Percentage frequencies of the three commonest gene arrangements in the third chromosomes of *Drosophila pseudo-obscura* in the Pinõn Flats area of Mount San Jacinto, California. Asterisks (*) mark the minimal annual frequencies of Standard (*ST*), and maximal frequencies of Arrowhead (*AR*) and Chirachua (*CH*) chromosomes (data of DOBZHANSKY, 1971).

Month and Year	ST	AR	CH	*n*	*Month and Year*	ST	AR	CH	*n*
Apr 1939	51	29	13	61	Sep 1952	58	19	14	104
May 1939	28*	36*	30	240	Feb 1953	61	19	6	444
Jun 1939	30	35	31*	154	Mar 1953	60	19	5	412
Aug 1939	36	33	26	156	Apr 1953	48	23	14	588
Sep 1939	51	23	23	190	May 1953	42	21	15	342
Oct 1939	55	25	17	284	Jun 1953	28*	37*	18*	60
Mar 1940	45	20	30	386	Nov 1953	70	13	2	180
Apr 1940	35	28	34	176	Feb 1954	61	18	5	164
May 1940	28	27	40	202	Mar 1954	56	19	6	86
Jun 1940	24*	30	42*	170	Apr 1954	56	23	5	304
Sep 1940	35	25	38	104	May 1954	44	27	11	128
Nov 1940	37	33*	26	80	Jun 1954	27*	41*	11	56
Mar 1941	56	11	24	110	Jul 1954	58	20	7	168
Apr 1941	58	20	17	110	Aug 1954	57	18	10	410
May 1941	44	28	24	100	Sep 1954	54	9	30*	106
Jun 1941	33*	32*	32*	192	Jul 1955	51*	15	18	192
Aug 1941	52	21	26	108	Aug 1955	58	22*	12	96
Sep 1941	56	19	16	80	Sep 1955	55	16	20*	146
Nov 1941	45	24	24	100	Oct 1955	57	22	11	250
Apr 1942	51	22	20	102	Apr 1956	47	26	12	248
May 1942	48	17	25	100	May 1956	37	30	16	228
Jun 1942	30*	23*	40*	114	Jun 1956	23*	40*	27*	52
Jul 1942	42	22	31	124	Jul 1956	49	18	20	228
Mar 1946	56	18	18	558	Aug 1956	57	17	15	108
Jun 1946	26*	24*	44*	500	Sep 1956	45	20	18	56
Mar 1952	49	16	19	74	Mar 1963	80	6	3	114
Apr 1952	51	20	21	156	Apr 1963	76	11	3	300
May 1952	30	23	33*	40	May 1963	63*	14*	5*	190
Jun 1952	26*	33*	29	140	Mar 1970	76	18	2	80
					Apr 1970	62*	22*	5*	1000

progressively over the same period. Finally at Berkeley the seasonal changes reverse those at Pinõn Flat and Andreas Canyon. The same is true of populations from Strawberry Canyon which is located at a height of 120 m among steep hills that form part of the Pacific Coast Range Mountains in the eastern part of the San Francisco Bay region. Moreover, here, rainfall shows a strong negative correlation with ST while temperature is well correlated with CH. This suggests that the adaptive advantages of these two arrangements may be associated with different environmental factors. Whether these factors are actually temperature and rainfall or whether undetermined selective factors are associated with these is not clear.

This variation in behaviour shown by different populations of *D. pseudoobscura* is not surprising when one considers the variety of climates and vegetation which occur within the distribution area of the species.

In *D. robusta* from Blackburg, Virginia, the inversions 2L–3 and XL–1 are high in overwintering flies but decline in spring and summer. These two arrangements are also especially frequent in northern and high altitude areas so that they may be connected with survival at low temperatures.

4·3 Hybrid zones

The Mexican lizard, *Sceloporus grammicus* consists of six parapatric populations differing from the standard female $X_1X_1X_2X_2$ *grammicus* karyotype of twelve metacentrics and twenty microchromosomes by a series of five centric fissions involving chromosomes 1 (polymorphic), 6 (fixed), 5 and 6 (both fixed), 5 (fixed) and multiple fissions (Fis_m) involving monomorphism for 2, 4, 5, 6 and one microchromosome and simultaneously polymorphic for fissions in 1 and 3. The male (X_1X_2Y) chromosome number in this species thus varies geographically from $2n=31$ to 46. Males with the standard karyotype are found over much of the range inhabited by the species. Although there appear to be no ecological or geographical barriers between many of the karyotypically distinctive populations, no evidence exists for sympatry or clinal intergradation. However, in the valley of Mexico several parapatric contacts, all involving very narrow zones of hybridization have been found between standard and Fis_6 near Cuernavaca, between S and Fis_m in the zone around the pyramids of Teotihuacan and between Fis_1 and Fis_6 near Amecameca and several sites near Rio Frio. The zone of hybridization between Fis_1 and Fis_6 on the south-west slopes of Cerro Potreto has been studied in some detail. Here the two populations meet along a broad front and interbreed in an area about 400 m wide, a distance about equal to the dispersal distance of an individual. Hybridization seems to be limited to the formation of F_1 and first backcross generations and there appears to be little introgression of Fis_1 and Fis_6 populations, which implies that they are being maintained in close geographic proximity despite an absence of effective pre-zygotic isolating mechanisms.

A comparable case is found in pocket gophers. *Thomomys umbrinus* ($2n=78$) has a high number of telocentrics plus three pairs of minute chromosomes and is limited to the Mexican plateau. *T. bottae* ($2n=76$) has a low number of telocentrics, no minutes and a more N. and W. distribution including most of the S.W. United States and N.W. Mexico. The two species are sympatric, or nearly so, in at least six localities in the north-western part of the range of *T. umbrinus*. In a sample from one such a locality in the Patagonia mountains a single F_1 individual was obtained with $2n=77$. The animal was a pregnant female with three nearly full term and apparently normal embryos. Three other specimens appeared to be backcross hybrids. Here too there is no apparent introgression and this, coupled with the narrowness of the hybrid zone, indicates restricted gene flow between the two species. Significantly where the two species are found in sympatry there is a marked ecological separation in the areas they occupy.

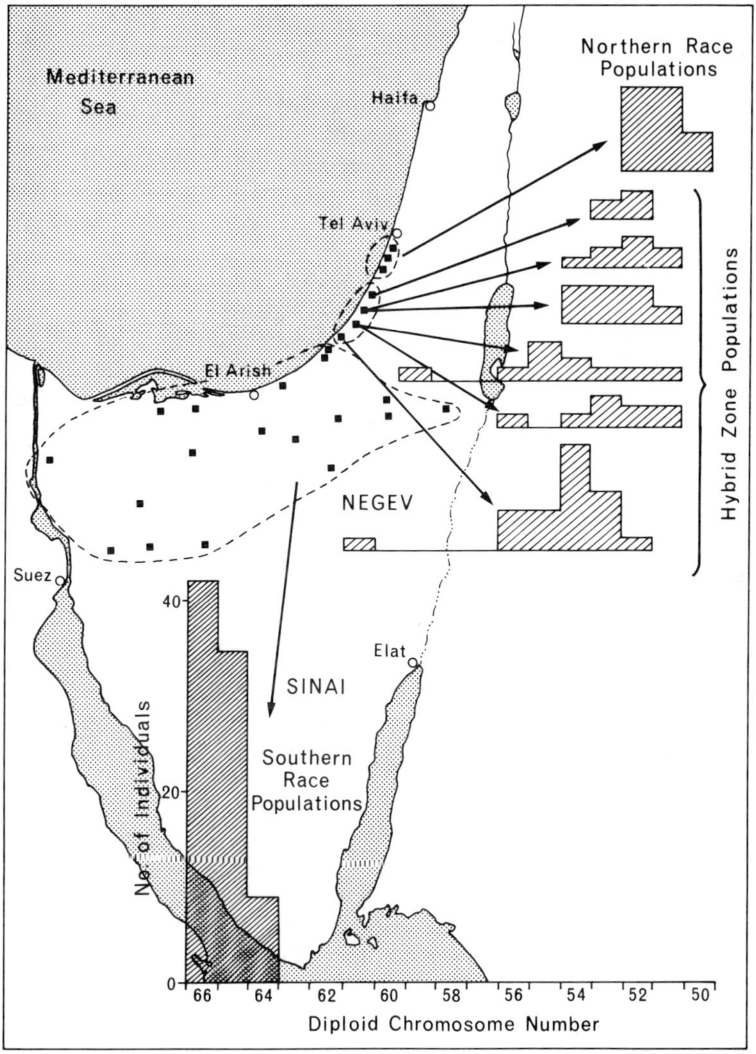

Fig. 4–4 Variation in chromosome number found in populations of the Greater Sand Gerbil, *Gerbillus pyramidum* in Israel and Sinai. (From WAHRMAN and GOUREVITZ, 1974, by permission of Israel Universities Press and Halsted Press—John Willey and Sons Inc., U.S.A.)

A unique hybrid zone has been reported in Israeli populations of the Greater Sand Gerbil, *Gerbillus pyramidum*, which exists in two major races differentiated by seven Robertsonian changes. The Southern race from the Sinai and the Negev possesses $2n=64-66$ while the much smaller northern race has $2n=50-52$. An extensive hybrid zone exists between the two races in which intermediate karyotypes of $2n=51-61$ are clinally distributed with the proportion of metacentrics increasing from north to south (Fig. 4–4).

These cases of hybrid zones, and others like them, are indicative of a confrontation between two formerly allopatric populations which have re-established secondary contact following divergence consequent upon separation. This is particularly well seen in the case of the *Ambysoma jeffersonianum* complex. Here there are two diploid bisexual species ($2n=2x=28$), a southern form (*A. laterale*), together with two triploid species ($2n=3x=42$) consisting entirely of females (*A. tremblayi* and *A. platineum*). The diploid species are largely allopatric but their ranges overlap in New York and New England and the triploids occur when the diploids approach one another or interdigitate. The two triploids consist of morphologically distinct populations between which no gene flow is possible and they too are largely allopatric. Even so each triploid occurs with one or another of the bisexual species because the unreduced triploid eggs will not develop unless activated by sperm produced by male diploids (gynogenesis). Thus *platineum* is found in association with *jeffersonianum* and *tremblayi* with *laterale* though triploid females can in fact mate with both kinds of diploid male. The triploids are very successful, have a wide range and usually form a large part of the population where they occur. Moreover, these diploid–triploid populations have been stable over at least 13–20 generations. Diploid males discriminate against triploid females and mate with them less frequently, accepting diploid females whenever there is a choice. Such preferential mating prevents triploid females from effectively swamping the diploid populations.

Likewise *Anolis aeneus* ($2n=34$) and *A. trinitatus* ($2n=36$) are common lizards on the island of Trinidad. These two species probably evolved in isolation and have made secondary contact in Trinidad as a result of human introduction. They differ in colour and in male territorial display. They tend to occur in pure species enclaves but since there are no ecological differences between them they overlap extensively. In areas of overlap they compete for mates since they have poorly developed premating isolating mechanisms and the resultant hybrids ($2n=35$) have greatly reduced reproductive fitness.

5 Chromosome Change and Evolutionary Change

Whom the gods love die young
No matter how long they live

Elbert Green Hubbard

5.1 Drift

Evolutionary change is currently viewed as resulting from the interplay between determinism and chance. The only chance event known is drift. Deterministic events include natural selection, a process of differential reproduction, and, according to some, genetic accumulation mechanisms.

There are several ways in which chance phenomena may influence cytogenetic processes in natural populations even when selection is operative. The most important of these is the effect of chance in influencing the transmission of chromosomes from one generation to the next. While such effects can occur in any population they become especially pronounced in small populations. Ernst Mayr has drawn attention to what is probably one of the most important forms of chance influence which he has termed the founder effect. This depends on the establishment of a new population by a few original founders—in the extreme case a single fertilized female—that carry only a small fraction of the total genetic variation of a parental population. If the founder population is successful it is bound to remain small for a number of generations so that founder events are inevitably followed by inbreeding.

The consequences of chromosome mutation can, therefore, be expected to differ in large and small populations. In large populations the behaviour of chromosome changes is expected to be influenced predominantly by selection, though compensation mechanisms and accumulation mechanisms may vitiate the effects of selection. In small populations, on the other hand, chromosome changes may be randomly eliminated or fixed as a result of inbreeding or drift. For example in the house mouse wild populations are sometimes subdivided into groups with an effective breeding size small enough to suggest that random drift, due to sampling error, and inbreeding may both be important factors governing population structure. Significantly populations of *Mus musculus* from isolated Alpine valleys have been shown to be homozygous for from 1–6 fusions.

Several possible cases of drift have also been reported in domestic animals where particular populations of cattle, pigs, goats and sheep have been found to carry autosomal fusions. In the SRB (Swedish Red and White) breed of cattle ($2n=60$) some 18% of the 2045 animals studied were heterozygous for a $1 \frown 29$ fusion involving the largest and the smallest autosome ($2n=59$). In addition a further 0.4% of these animals were fusion homozygotes ($2n=58$). The relatively high frequency of this translocation has been explained on the basis of genetic drift. Thus the effective population size in cattle is small since breeds have been subdivided into isolated local populations in which males are very limited in number compared with females. In addition, sires are recruited principally from certain select herds within which line breeding is often practised. Under such circumstances an individual heterozygous bull can have considerable influence on the genetic constitution of the population.

In the case of SRB cattle, as indeed in similar cases in pigs and goats, there is no evidence for either male or female infertility. In the sheep, on the other hand, the fusion was detected because it led to sterility. Two such cases are on record; both occur in elite breeds from New Zealand. In one of these breeds, the Romney, 5% of a sample of 180 sheep carried a heterozygous fusion between autosomes 5 and 26 ($2n=53$). In some such heterozygotes there was complete sterility and a cessation of spermatogenesis in late primary spermatocytes. This cannot apply to all of them since the condition is clearly transmissible. In this flock, close breeding has been followed for over 20 years and of 6 relatives of the propositus examined 4 were carriers of the translocation, including the parent dam, the twin sister of the propositus and his two male half sibs.

A much higher incidence (26.3% in a total of 327 sheep) of a morphologically similar fusion between autosomes 7 and 25 occurs in the elite flocks of the Drysdale breed (Table 10) where both structural

Table 10 Incidence of fusion heterozygotes and homozygotes in four flocks of Drysdale sheep from New Zealand (data of BRUÈRE *et al.*, 1972).

	Chromosome number								*Combined incidence (%)*	
	Rams				*Ewes*				$2n=53$	$2n=52$
Flock	54	53	52	Total	54	53	52	Total		
Massey	16	4	2	22	100	23	1	124	18.5	2.1
Utawai	21	7	1	29	44	25	2	71	32.0	3.0
Pahiatu	17	8	0	25	19	5	0	24	26.5	0.0
Pohangina	22	4	0	26	—	—	—	—	15.4	—
Totals	76	23	3	102	163	53	3	219	23.7	1.9
	321								26.3	

homozygotes ($2n=52$; 2%) and heterozygotes ($2n=53$; 24%) were present in addition to basic homozygotes ($2=54$). In the Drysdale flock neither the rams nor the ewes appeared to be affected in terms of genitalia or fertility. In this breed four stud lines are responsible for all the rams used in multiplying flocks. The breed had its origin in 1931 and, from the available pedigree data, it is clear that the fusion must have been in the breed for a number of years and it can be traced back positively to 1962. Significantly too the Romney breed was used in the development of Drysdale sheep.

These cases suggest that a chromosome mutation with a weak negative effect may become widely spread within a short time under the breeding conditions prevalent in domestic animals. Chance effects may have a considerable influence in such populations which are restricted in size and subdivided into small groups in which the relatively few breeding animals are of varying reproductive capacity. However, while it is possible for a rearrangement of indifferent survival value to gain a foothold in a single small population through drift it is highly improbable that a rearrangement introduced by drift would become established in a large population and even less probable that the same rearrangement would come to characterize different populations of a given species by this means.

5·2 Mechanisms of accumulation

A second school of thought argues that some at least of the chromosome changes found in nature occur because they have efficient mechanisms of accumulation. Nur, in particular, has been a consistent advocate of the view that accumulation mechanisms serve to maintain deleterious B-chromosomes at equilibrium frequencies in natural populations. The most forceful of the cases which appears to support his view is that which he has presented for the mealy bug *Pseudococcus obscurus* where, as we have seen earlier (p. 33), supernumerary chromosomes pass preferentially and without reduction into functional male gametes. Under laboratory conditions it has been shown that the developmental rate of B-containing males is decreased, especially at low temperatures, as too is the total number of sperm produced by such males. In consequence Nur argues that the proportion of B-chromosomes in a population reaches a stable equilibrium which is related to the extent to which accumulation is able to overcome the decrease in fitness produced by the B's. Similar claims have been made for other species too.

There are, however, several qualifications that need to be made to the *Pseudococcus* story which indicate that the situation is more complicated than a simple balance between accumulation and negative selection.

(i) In matings in the wild, transmission values are often much lower than those obtained in the laboratory. This is attributed to multiple mating of one female with different males but no clear demonstration of this has been provided.
(ii) B's have no known effect on female viability or fecundity.
(iii) The transmission frequency varies in different males (Table 11).

Table 11 | Summary of the rates of transmission (k values) of males with various numbers of B-chromosomes in *Pseudococcus obscurus*. *N* is the number of males assigned to each class. The mean *k* of each class and the standard error are based on the actual values (data of NUR, 1966).

	% males with			
k range	1B	2B	3B	4B
0.01–0.09	1.3			
0.10–0.19		2.8	3.4	12.5
0.20–0.29	1.3	5.6		12.5
0.30–0.39	2.6	4.2	17.2	
0.40–0.49	2.6	8.4	6.9	37.5
0.50–0.59	6.5	11.3	13.8	
0.60–0.69	3.8	11.3	20.7	12.5
0.70–0.79	11.7	8.4	17.2	12.5
0.80–0.89	10.4	11.3	10.4	
0.90–0.99	26.0	26.8	10.4	12.5
1.00	33.8[a]	9.9		
N	77	71	29	8
Mean $k(\bar{k})$	0.837±0.024	0.708±0.026	0.631±0.039	0.412±0.085

[a] includes one male with $k = 1.05$

81.6% of males with single supernumeraries had transmission values in excess of 0.5 but the mean transmission value decreases with increasing B-number.

(iv) The number of B's present in females shows seasonal fluctuations being higher in the summer and fall than in the late winter and spring. This, in turn, suggests that the relative fitness of 0B males may change seasonally and that the great advantage of 0B males over 1B males may be restricted to only part of the year.

(v) The number of B's present per population varies greatly. Frequencies are high in coastal collections (range 1.21–2.74 B's per individual) while those inland are low (range 0.01–0.92 per individual). This suggests that B frequencies are affected by different ecological conditions.

Likewise in grasshoppers where, as we have seen (p. 32), there is clear evidence for preferential transmission of B's at female meiosis there are also indications that the situation is complex rather than simple.

(i) In *Melanoplus* there are two types of B-chromosome, a medium sized metacentric and a small telocentric. Only the former shows preferential transmission. Thus the high rate of transmission of the metacentric B is a property of the B-chromosome itself and not all B's are subject to meiotic drive.

(ii) Although the overall ratio of male/female embryos obtained from the *Melanoplus* crosses using the metacentric B was 1 : 1 there was significantly more 0B female than 0B male and correspondingly more 1B male than 1B female progeny. Since the X and B univalents move randomly at male meiosis and since sex is regulated by the X-chromosome content of the sperm these inequalities in the distribution of the two sexes must involve a differential utilization of the four types of sperm—X, X+B, B and nullo—produced by B-containing males. How this operates is not known.

(iii) In *Myrmeleotettix* it is clear from the transmission studies that the behaviour of the B depends on its origin. That from Tal-y-Bont shows preferential transmission while that from E. Anglia does not (Table 5) and this despite the fact that both populations maintain a high B-frequency (E. Anglia 60%, Tal-y-Bont 70%) which has been stable over 3–5 years.

(iv) While there is no inequality in the distribution of the B's in male and female progeny of the type found in *Melanoplus* there is a significant loss of B's when transmitted by the East Anglia males. There is no simple explanation for such a loss in terms of the behaviour of the B at male meiosis and it must occur at a post-meiotic stage, that is either during spermiogenesis or at fertilization. There is a similar unexplained loss of B's in the male in *Chloealtis conspersa* where the joint data (Table 4) suggest that preferential transmission in the female more than balances loss in the males.

In summary, there are clear differences in the transmission rates of B-chromosomes in both male and female according to their origin and their genetic structure. This implies that different B's may be maintained in different populations by different combinations of the same controlling factors and that even the same B, at least in morphological terms, may behave differently in different populations and against different genetic backgrounds.

Some have concluded that to be permanently maintained within a population any supernumerary system must be endowed with an accumulation mechanism from the outset so that it is protected from loss by either selection or drift. It may well be that meiotic drive and other boosting mechanisms are necessary to allow what must initially be at least partial trisomic states to increase rapidly in frequency when introduced into a local population. But whether such mechanisms alone are capable of maintaining such a system once it is established is open to dispute. Even in *Lilium callosum*, where preferential segregation is held to play a key role in determining the B-frequency of a population, striking differences are in fact found between mean B-frequencies in wild populations. One explanation for such differences is that populations possess different equilibria under different ecological conditions. It appears much more likely, therefore, that the B-chromosome frequency of a population is in fact determined by a variety of interacting components including differential inheritance leading to both gain and loss and differential selection both positive and negative.

Inherent in the assumption that accumulation mechanisms are essential for the maintenance of B-chromosomes is the belief that because B's have no demonstrable positive effects on the adaptedness of the individuals that carry them then they have no useful genetic effects of any kind. A distinction can, however, be drawn between the advantage of possessing an adaptive genotype and that of transmitting adaptive genotypes. The latter is not simply a corollary of the former. Thus the fact that chromosome mutations do not improve the adaptedness of their possessors in no sense excludes them from influencing the cytogenetic composition of the gametes and zygotes produced by an individual and hence subsequently influencing the genetic composition of its progeny.

For example several supernumerary systems have been shown to modify the pattern of chiasma formation in individuals where they are present. The precise effect varies from species to species but the net effect is the same in all of them—the supernumerary is capable of producing a wider range of genotypes among the progeny of B-containing individuals. In rye while the slight reduction in mean cell chiasma frequency with increasing B-number is not significant there are positive correlations between the B-chromosome frequency and the variation between pollen mother cells within plants (between cell variance) on the one hand (Fig. 5–1) and an increase in the asymmetry of chiasma distribution between chromosome arms. The chiasma frequency variation between plants also increases with increasing B-number (Table 12). In keeping with these several effects it has been shown that the presence of B's in parent plants increases the phenotypic variability of

Table 12 Relationship between B-chromosome content, mean cell chiasma frequency and mean plant chiasma variance in Rye (data of JONES and REES, 1967).

B-class	0B	1B	2B	3B	4B	5B	6B	7B	8B
Mean cell chiasma frequency	14.85	13.43	13.22	13.39	12.64	12.86	13.20	12.64	13.03
Mean plant chiasma variance	1.270	4.651	1.276	1.808	1.047	2.132	2.324	9.702	3.724

their progenies. Likewise in *Zea mays* mean cell chiasma frequency increases with increasing B-chromosome frequency (Fig. 5–1) and regression analysis shows that the increase is significant.

The chiasma is a visible expression of genetical crossing-over and, as such, serves as a potential marker of the extent of recombination between linked genes. The control exercised by B-chromosomes on chiasma frequency variance and distribution thus provides a mechanism for

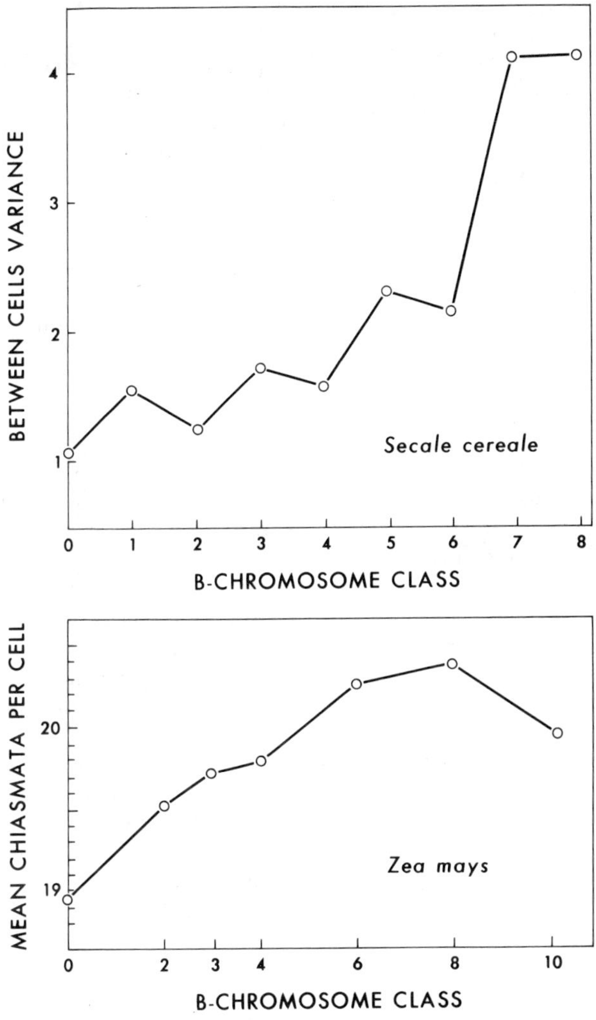

Fig. 5–1 The effect of B-chromosomes on chiasma characteristics in two plant species. In rye mean cell chiasma frequency is not affected (Table 13) but between cell variance is. Additionally the variation in chiasma frequency between pollen mother cells is disproportionately high in the presence of odd, as compared with even, numbers of B-chromosomes. Covariance analysis indicates that the mean variances of the odd and even classes is significantly different. The B-chromosome thus exerts a differential influence depending on whether it is present in odd or even numbers (compare with Table 12). In maize mean cell chiasma frequency increases with an increasing B-chromosome number and regression analysis indicates the increase is statistically significant. (Upper part from JONES and REES, 1967; lower part from AYONOADU and REES, 1968.)

promoting extra variability in diploid populations. It is true of course that the evolutionary importance of chiasma frequency and chiasma variance can be assessed only in relation to all the other aspects of the recombination system. It is also true that chiasma frequency *per se* is a relatively insensitive estimator of the variability generated by meiosis since one has no accurate idea of what genetic loci are being recombined. Nevertheless, these examples emphasize the principle that the effects of chromosome mutations may well operate through the meiotic system rather than the epigenetic system. And equivalent effects on chiasma distribution and frequency are known for supernumerary segment systems, for fusion/fission systems and for inversions (Table 7 and p.74).

5·3 The cytogenetics of speciation

Chromosome differences occur not only within populations and between populations but also between races within species and between species. Thus in hybrids between races or species the chromosomes may fail to pair or else may pair not as bivalents but as multiples or as multivalents because the parental forms differ by fixed structural or numerical changes. This, in turn, leads to lowered fertility or in extreme cases to sterility because of the production of aneuploid gametes.

Where there is a parallelism between interspecific cytogenetic differences and intraspecific chromosomal polymorphisms it is commonly assumed that the former are relics of a previously existing polymorphic situation which has collapsed to give rise to a series of geographic isolates homozygous for different chromosome sequences. In *Drosophila*, as Bock has pointed out, the inversion differences distinguishing members of a species complex or group are sometimes substantially different in type from those detected as extant polymorphisms within the species concerned. Coupled with this, members of a species group which exhibit little inversion polymorphism in natural populations show relatively few differences in gene arrangement in interspecific hybrids whereas highly polymorphic species generally show extensive differences in species hybrids. These facts have led to the suggestion that there has been a cyclical process of fixation of inversions, coincident with the formation of a new species, followed by the origin of new inversions within these species which then provide a basis for new floating polymorphisms.

Whether or not this argument is valid there can be no doubt that in some cases the kinds of chromosome differences which distinguish species are quite different from those which occur as polymorphisms within species. Thus in *Clarkia* the translocation heterozygotes found in wild populations, or those derived from inter-population crosses within a species, show catenations which involve all of the interchanged chromosomes and give a regular alternate disjunction. By contrast those which distinguish different species behave very differently as heterozygotes. Thus maximum multiples are infrequent and when present they do not disjoin in a regular manner. Those translocations that differentiate species, therefore, are structurally of a quite different type from those found within species. Those that persist as variable elements within populations involve the exchange of large segments or entire chromosome arms while those which differentiate species have originated as short or unequal exchanges. Indeed there are reasonable grounds in *Clarkia* for concluding that chromosome repatterning has played an integral part in the speciation process. Such repatterning can be expected to lead to the rapid development of genetic barriers to inter-breeding, barriers moreover which a species can take with it wherever it moves or migrates. In such cases it is clear that new monomorphisms come not from the collapse of stable balanced polymorphisms but rather from short-lived transient polymorphic states which reach rapid fixation.

Where isolation is due to chromosomal change acting at a postmating level it is necessary to account for the initial establishment of a chromosome rearrangement which leads to a reduced fertility in the heterozygous state. Since structural rearrangements originate as single events in the heterozygous form this implies that, in a large panmictic population, structural homozygotes cannot be produced for several generations, and even then would occur only in small numbers. It is generally believed, therefore, that rearrangements which lower fecundity in the heterozygous stage can only reach fixation in a population of small size. Thus reduced competition following a reduction in population size could provide a mechanism for transporting chromosomal changes through the bottleneck of sterility which accompanies the inception of many chromosomal mutations.

Alternatively some authors have claimed that the occurrence of chromosome differences between races or species may indicate not that these differences have played a causative role in speciation but rather that the circumstances under which species arise, involving small local and perhaps strongly isolated populations, may be particularly favourable for the establishment and fixation of certain types of rearrangement. While this may be true of some cases it cannot apply to others. Thus the translocations which differentiate the species of *Clarkia* and which behave irregularly at meiosis in hybrids certainly serve as reproductive isolating agents.

In the *viatica* group of morabine grasshoppers the 'coastal' forms have strictly parapatric distributions with narrow zones of overlap in which hybrids may be found at least in some cases. These parapatric taxa differ cytogenetically in respect of chromosome rearrangements, for which they are homozygous. The narrowness of the zones of overlap is a reflection of the reduced fecundity and perhaps also in some cases the diminished viability of the hybrids. The presence of these hybrids indicates that if premating isolation exists at all in these cases it is weak and it is difficult to avoid the conclusion that the chromosomal rearrangements which distinguish the taxa have played a primary role in generating isolating mechanisms in this group too.

As currently conceived there are three possible modes of primary speciation. In the first of these, isolation develops fortuitously in physically separated populations as a by-product of evolutionary divergence determined either by drift or by directional selection. In the two other types reproductive isolation results from selection against hybridization. In one of these, divergence is allopatric in origin but is reinforced by stabilizing selection which operates to reduce the reproductive wastage consequent upon the lowered fitness of the hybrid progeny generated when the two allopatric populations meet again secondarily. Alternatively both divergence and isolation may occur under sympatric situations as a result of disruptive selection. Sympatric speciation thus implies that new races and new species arise within the dispersal range of the parental population. Polyploidy is without question the most typical case of abrupt and sympatric speciation for, at its origin, a polyploid is likely to be the only one of its kind amongst a population of diploids (autopolyploid) or, at best, the new polyploid population will consist of a small number of individuals only (allopolyploid).

6 Population Genetics and Population Cytogenetics

You mentioned your name as if I should recognize it, but beyond the obvious fact that you are a bachelor, a solicitor, a freemason and an asthmatic I know nothing whatsoever about you.

Sherlock Holmes

Most of the standard texts on population genetics have concentrated on allelic variation in natural populations. Indeed theoretical developments in this area have been formulated almost exclusively in terms of single or simple gene systems which obey the law of independent segregation. The genes within a population are, however, permanently held and dispersed as a set of particular genotypes organized into a system of linkage units or chromosomes. Moreover, within individual chromosomes, genes do not exist against random genetic backgrounds. On the contrary, correlated gene blocks are probably the rule rather than the exception. Few adaptations have been found to involve single gene effects so that mechanisms must exist to provide for and promote the development of interacting systems of non-allelic genes. The properties of any given gene locus are thus determined by the effects of other loci which form a linkage association with it.

For example, single heterozygous inversions lead to the production of a gene complex which is protected against dissociation by recombination and so allows for the secondary development of epistatic (non-allelic) interactions between such linked loci. But linkage association through inversions can be carried even further. Some independent inversions on the same chromosome tend to consistently occur together even when the possibility of recombination exists between them. That is, effective recombination between such inversions is much lower than would be expected on the basis of the genetic distance between them. Such an adaptively interacting system of linkage associations is sometimes referred to as linkage disequilibrium. In *Drosophila subobscura* inversion polymorphism is known in all five chromosomes of the haploid complement and a total of 61 inversions are found with the following distribution pattern:

Chromosome					*Total inversions*
A(=X)	E	I	O	U	
5	10	3	29	14	61

While the different interchromosomal karyotypic combinations seem to occur at random, linkage disequilibrium between independent inversions of the same chromosome is in many cases complete in this species. The gene arrangements on the two arms of the X-chromosome of *D. robusta* also provide evidence for such an interacting system and there are at least three different X-chromosome disequilibrium systems.

Inversion sequences demonstrate a further feature of importance. When heterozygous they are equally potent boosters of recombination elsewhere in the complement as they are of conservers of recombination within the inverted segments themselves. In *Drosophila*, for example, inversions in the X-chromosome and in chromosome-II increase crossing over in chromosome-III. While genotypically controlled chiasma localization can lead to a restriction of recombination it does not produce equivalent inter-chromosomal effects.

It is widely acknowledged that sexual reproduction and genetic recombination are capable of generating a variety of genotypes. But allelic heteroxygosity, or indeed homozygosity too, is destroyed and reassembled in every generation by the normal events of meiosis and fertilization. The virtue of at least some chromosome changes in evolution is that they ensure that specific allelic associations can be passed unaltered from generation to generation. Thus although allopolyploidy determines a change to greater inbreeding it nevertheless conserves heterozygosity. The same is true of interchange heterozygotes. Again while the major advantage of female parthenogenesis (thelytoky) is the increase it offers in the reproductive potential of a population due to the exclusive production of females, an additional advantage conferred especially by obligate apomitic thelytoky is the ability to perpetuate adaptive genotypes in an unchanged form generation after generation. Thus chromosome mutations can both create and conserve adaptive allelic associations.

Of course where chromosome mutants occur at a low frequency in one or more populations of a species, as for example in the human cases cited earlier (p. 12), they are probably transients without adaptive or evolutionary importance. These are maintained either by recurrent mutation or else by hereditary transmission for not all chromosome mutants arise *de novo*, some are familial, the mutant being transmitted by one of the parents (Table 13). Alternatively inbreeding, with or without drift, may cause a particular mutation to spread through a small mating group as in the case of cattle and sheep (p. 64). Where, on the other hand, mutants are frequent, have a wide geographical range and show either temporal stability or else predictable variability there are clear grounds for considering an adaptive role. Even so, in most of the cases described in this book, or indeed most of the cases that have ever been described, we do not know in detail how the observed patterns of cytogenetic variation have been created or are currently being maintained. Nor is this surprising. In only a few organisms do we have any adequate knowledge about their gene content let alone the possibilities for heterotic or epistatic interaction. Neither do we have an adequate understanding of the historical and biogeographical events which have led to the present day patterns of distribution within and between species. Yet there can be little doubt that the evolution of many present day species is linked with the dynamic changes of both physical and biotic conditions which occurred during and since late pleistocene times.

Table 13 Frequencies of balanced rearrangements in 42 human babies where both parents have also been examined (data of Jacobs).

		Both parents		*Altered parent*	
Type of rearrangement	*No. of cases*	*Examined*	*Normal*	♂	♀
Whole arm transpositions (Robertsonian translocation)					
D/D	32	12	1	5	6
D/G	9	6	1	2	3
Reciprocal translocations	41	20	7	6	7
Inversions	7	4	0	2	2
Totals	89	42	9*	15	18
				33†	

N.B. * These 9 cases represent new mutants so that 21% of euploid rearrangements reported here are new.
† Since 33 of the 42 cases examined gave evidence of familial transmission this gives an estimate of 70% for transmitted mutants.

Most of our evidence relating to the function of cytogenetic variation in natural populations is thus based on correlation and inference. In some instances fairly clear cut correlations exist between cytogenetic structure and ecology. For example in the four best known instances of aneuploid reduction in plants—*Crepis*, *Haplopappus*, *Chaenactis* and *Clarkia*—the reduction series has developed in conjunction with a xeric habit. Again in the midge *Kiefferulus intertinctus* the Lonsdale (*Lo*), Barwon (*Ba*) and Corio (*Co*) inversion sequences all increase the length of the ventral tubules in both male and female larvae. Larger ventral tubules provide an increased area for respiratory exchange. Significantly the Lo and Ba inversions show a non-random association (p. 71) and are more common in warmer regions where the oxygen content of the water is low. In others no such simple correlates exist and a complex of inter-related factors must be involved.

The major outstanding problems in population cytogenetics are thus twofold:

(i) while the causative basis for cytogenetic change in natural populations must relate to the mode of life of the organism there is a lack of comprehensive studies which offer combined information on the distribution, the ecology, the historical biogeography and the temporal characteristics of populations, and

(ii) while there can be no doubt that chromosome changes serve to regulate the pattern of genetic variability in natural populations through alterations in the linkage and recombination systems (Fig. 6–1) there is a lack of detailed understanding of the effects of chromosome change on the genetic systems of populations.

This book will, therefore, have served its purpose if it provides a stimulus for at least some of those who read it to contribute to the eventual resolution of these areas of ignorance.

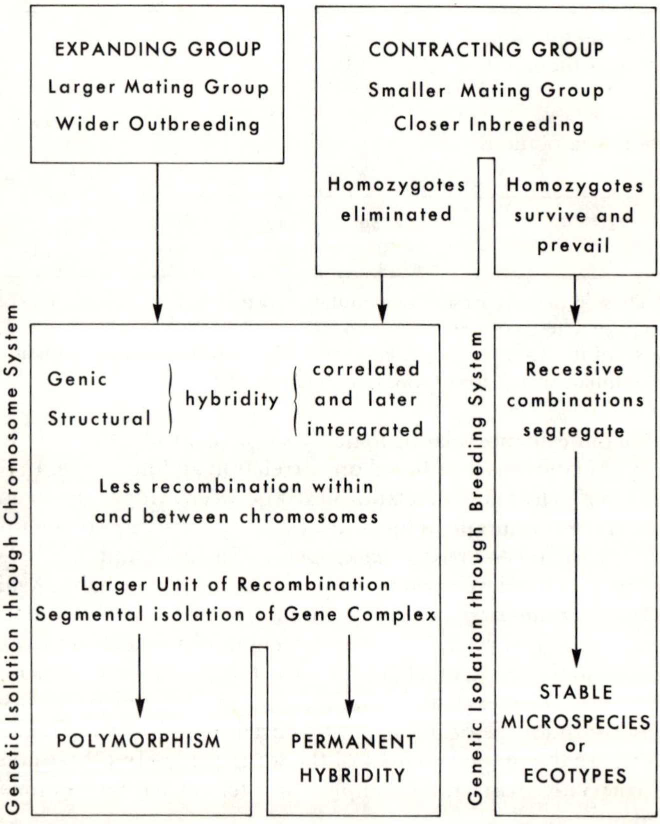

Fig. 6–1 Differential patterns of change in the genetic system resulting from alterations in the external and genetic environments of natural populations. (From DARLINGTON, 1956, by permission of the Royal Society.)

Further Reading and References

There is an art of reading, as well as an art of thinking, and an art of writing.
Isaac D'Israeli

General

DARLINGTON, C. D. (1974). *Chromosome Botany*. 3rd edn. Allen and Unwin, London.

JOHN, B. and LEWIS, K. R. (1975). *Chromosome Hierarchy*. Oxford University Press, London.

MAYNARD SMITH, J. (1958). *The Theory of Evolution*. 2nd edn. Penguin Books, Harmondsworth.

STEBBINS, G. L. (1971). *Chromosomal Evolution in Higher Plants*. Edward Arnold, London.

Chapter 1

DARLINGTON, C. D. (1956). *Proc. Roy. Soc., B*. **145**, 350–64.

GRANT, V. (1958). *Cold Spring Harbor Symp. Quant. Biol.*, **23**, 337–63.

Chapter 2

CARSON, H. L. (1946). *Genetics*, **31**, 95–113.

CHANDLEY, A. C., CHRISTIE, S., FLETCHER, J., FRACKIEWICZ, A., and JACOBS, P. A. (1972). *Cytogenetics*, **11**, 516–33.

EVANS, H. J. (1973). *Brit. med. Bull.*, **29**, 196–201.

FORD, C. E. and EVANS, E. P. (1974). *Chromosomes Today*, **4**, 387–97.

FREDGA, K. and BERGSTROM, U. (1970). *Hereditas*, **66**, 145–52.

PATHAK, S., HSU, T. C. and ARRIGHI, F. E. (1973). *Cyto. Cell. Genet.*, **12**, 315–26.

STRID, A. (1968). *Bot. Notis.*, **121**, 153–64.

TETTENBORN, U. and GROPP, A. (1970). *Cytogenetics*, **9**, 272–83.

Chapter 3

BOSEMARK, N. O. (1967). *Hereditas*, **57**, 239–62.

CLELAND, R. (1972). *Oenothera Cytogenetics and Evolution*. Academic Press, London.

FORD, C. E. and HAMERTON, J. L. (1970). *Symp. Zool. Soc. Lond.*, **26**, 223–36.

FROST, S. (1958). *Hereditas*, **44**, 75–111.

GALLAGHER, A., HEWITT, G. M. and GIBSON, I. (1973). *Chromosoma*, **40**, 167–72.

HEWITT, G. M. (1973). *Chromosoma*, **40**, 83–106.

HEWITT, G. M. and JOHN, B. (1967). *Chromosoma*, **21**, 140 62.

HEWITT, G. M. and JOHN, B. (1968). *Chromosoma*, **25**, 319–42.

JAMES, S. H. (1970). *Heredity*, **25**, 53–77.

JOHN, B. and LEWIS, K. R. (1958). *Heredity*, **12**, 185–97.

JOHN, B. and QURAISHI, H. B. (1964). *Heredity*, **19**, 147–56.

JONES, K. (1974). *Chromosoma*, **45**, 353–68.

KAYANO, H. (1971). *Heredity*, **27**, 119–23.

LOWE, C. H., WRIGHT, J. W., COLE, CH. J. and BEZY, R. L. (1970). *System. Zool.*, **19**, 128–41.

LUCOV, Z. and NUR, U. (1973). *Chromosoma*, **43**, 289–306.

LEWIS, W. H., OLIVER, R. L. and SUDA, Y. (1967). *Ann. Missouri Bot. Garden*, **54**, 153–71.

NAGAMATSU, T. and NODA, S. (1971). *Cytologia*, **36**, 332–40.

NUR, U. (1966). *Genetics*, **54**, 1225–38.

NUR, U. (1969). *Chromosoma*, **27**, 1–9.

SNOW, R. (1964). *Genetica*, **35**, 205–35.
THIEN, L. B. (1969). *Evolution*, **23**, 456–65.
UESHIMA, N. (1967). *Chromosoma*, **20**, 311–31.
VOSA, C. G. (1973). *Chromosoma*, **43**, 269–78.
YOSIDA, T. H., TSUCHIYA, K. and MORIWAKA, K. (1971). *Chromosoma*, **33**, 252–67.

Chapter 4

BANTOCK, C. R. and COCKAYNE, W. C. (1975). *Heredity*, **34**, 231–45.
BRADSHAW, W. N. and HSU, T. C. (1972). *Cytogenetics*, **11**, 436–51.
BRNCIC, D. (1962). *Chromosoma*, **13**, 183–95.
CARSON, H. L. (1958). *Cold Spring Harbor Symp. Quant. Biol.*, **23**, 291–305.
DA CUNHA, A. B., DOBZHANSKY, TH., PAVLOVSKY, O. and SPASSKY, B. (1959). *Evolution*, **13**, 389–404.
DOBZHANSKY, TH. (1971). In *Ecological Genetics and Evolution* (ed. Creed, R.) Blackwell, Oxford. Pp. 109–33.
GORMAN, G., LIGHT, P., DESSAUER, H. C. and BOOS, J. O. (1971). *Systematic Zool.*, **20**, 1–18.
HALL, W. P. and SELANDER, R. K. (1973). *Evolution*, **27**, 226–42.
HEWITT, G. M. and JOHN, B. (1970). *Evolution*, **24**, 169–80.
PATTON, J. L. and DINGMAN, R. E. (1968). *J. Mammal.*, **49**, 1–13.
PATTON, J. L. (1970). *Chromosoma*, **31**, 41–50.
SCHROETER, G. (1968). Pericentric inversion polymorphism in *Trimerotropis helferi* and its effect on chiasma frequency. Ph.D. Thesis Univ. California Davis (69–824 Univ. Microfilms Inc., Ann Arbor, Michigan, USA).
STAIGER, H. (1954). *Chromosoma*, **6**, 419–78.
YOSIDA, T. H., TSUCHIYA, K. and MORIWAKA, K. (1971). *Chromosoma*, **33**, 30–40.
WAHRMAN, J. and GOUREVITZ, P. (1974). *Chromosomes Today*, **4**, 399–424.

Chapter 5

AYONOADU, U. and REES, H. (1968). *Genetica*, **39**, 75–81.
BOCK, I. R. (1971). *Chromosoma*, **34**, 206–29.
BRUÈRE, A. N. (1969). *Cytogenetics*, **8**, 209–18.
BRUÈRE, A. N., CHAPMAN, H. M. and WYLLIE, D. R. (1972). *Cytogenetics*, **11**, 223–47.
HEWITT, G. M. and BROWN, F. M. (1970). *Heredity*, **25**, 363–71.
JONES, R. N. and REES, H. (1967). *Heredity*, **22**, 333–47.
KAYANO, H. (1962). *Evolution*, **16**, 246–53.
LEWIS, H. (1962). *Evolution*, **16**, 257–71.
NUR, U. (1966). *Genetics*, **54**, 1239–49.
REES, H. and HUTCHINSON, J. (1974). *Cold Spring Harbor Symp. Quant. Biol.*, **38**, 174–82.
WAHRMAN, J. (1954). *Proc. 9th Int. Congress Genetics. Caryologia Suppl.* 1954, 683–4.
WHITE, M. J. D. (1970). *J. Aust. Ent. Soc.*, **9**, 1–6.

Chapter 6

JOHN, B. and LEWIS, K. R. (1966). *Science*, **152**, 711–21.
LEVITAN, M. (1973). *Evolution*, **27**, 476–85.
LEWIS, H. (1953). *Evolution*, **7**, 102–9.
MARTIN, J. (1973). *Aust. J. Biol. Sci.*, **26**, 1371–7.
SPERLICH, D. and FEUERBACH MRAVLAG, H. (1974). *Evolution*, **28**, 67–75.

UNIVERSITY OF BRISTOL
LIBRARY
22 JUN 2024
WITHDRAWN

UNIVERSITY
OF BRISTOL
VET. SCIENCE

77 07765